Electronics for Technicians 3

Electronics for Technicians 3

S. A. Knight BSc(Hons)Lond

Senior Lecturer in Mathematics & Electronic Engineering,
Bedford College of Higher Education.

THE BUTTERWORTH GROUP

UNITED KINGDOM Butterworth & Co. (Publishers) Ltd
London: 88 Kingsway, WC2B 6AB

AUSTRALIA Butterworths Pty Ltd
Sydney: 586 Pacific Highway, Chatswood, NSW 2067
Also at Melbourne, Brisbane, Adelaide and Perth

CANADA Butterworth & Co. (Canada) Ltd
Toronto: 2265 Midland Avenue, Scarborough, Ontario M1P 4S1

NEW ZEALAND Butterworths of New Zealand Ltd
Wellington: T & W Young Building, 77–85 Customhouse Quay, 1, CPO Box 472

SOUTH AFRICA Butterworth & Co. (South Africa) (Pty) Ltd
Durban: 152–154 Gale Street

USA Butterworth (Publishers) Inc
Boston: 10 Tower Office Park, Woburn, Mass. 01801

First published 1980

British Library Cataloguing in Publication Data

Knight, Stephen Alfred
Electronics for technicians. – (Technician Education Council. TEC technician series).
3
1. Electronic apparatus and appliances
I. Title II. Series
621.381 TK7870 79-41649

ISBN 0-408-00458-4

Typeset by Reproduction Drawings Ltd, Sutton, Surrey

Printed in England by Page Bros Ltd., Norwich

Preface

This book covers the course requirements for the TEC Electronics 3 Unit 76/009. This is an essential unit which is studied in the third and final certificate year of the electronics and telecommunications technicians programme. Much of the content of other options such as light current applications are also covered in the text. Certain 'non-essential' parts of the syllabus have been deliberately omitted, but these can normally be filled in by the lecturer strictly as additional notes.

The order in which the book is written does not follow the order of the guide syllabus as it is not possible, or at least it is difficult, to go through the work in a manner that represents a logical progress if the syllabus order is adhered to rigidly. For example, the effect of the removal of the source (or emitter) bypass capacitor on an amplifier's performance is best dealt with under the heading of negative feedback, and field-effect devices are probably best discussed after bipolar transistors if the most useful comparisons are to be made. Consequently, the book begins with a simple treatment of pulse shaping passive circuits and proceeds then to bipolar transistor small-signal amplifiers before field-effect transistors are considered. Noise and negative feedback then precede the section on large signal amplifiers. Oscillators and waveform generators are next covered, followed by stabilised power supplies.

To keep the book to a reasonable length notes on integrated circuits unfortunately have had to be omitted. However, what is actually contained in a miniature form in an integrated circuit are circuits of the form discussed throughout the rest of the book. Consequently because the reader happens to be a technician who will use and work with integrated circuits and not design them, an understanding of how amplifiers work and the effect of negative feedback on amplifier performance will make it an easier matter for the student to appreciate the properties of integrated systems and the way these properties are interrelated in the designer's final product when he comes to study the fabrication of such packages.

The reader should have already completed the course requirements of TEC Unit 76/001 for Electronics 2. A knowledge of the fundamental properties of bipolar transistors as both amplifiers and oscillators has been assumed.

As in the other books for TEC courses—*Electrical Principles for Technicians 2* and *Electronics for Technicians 2*—a large number of worked examples have been included throughout the text and self-assignment problems will be found not only in the text but at the end of each section. All problems have their solutions given at the end of the book and are numbered throughout each section in order so that no difficulty arises in checking the answers. Many of the solutions contain additional notes or comments to clarify certain points which might otherwise be confusing or point the way to a method by which that solution has been obtained.

Acknowledgement is made to the City & Guilds Institute for permission to use a number of their past examination questions. The solutions given to these problems are the sole responsibility of the author.

S. A. Knight

Contents

1 Resistive-capacitative networks

Aims: At the end of this Unit section you should be able to:
Sketch and label a rectangular wave showing pulse duration, pulse amplitude, rise-time and decay-time.
Understand the basic processes in the charging of a capacitor.
Know what is meant by time-constant CR.
Understand the operation of differentiating and integrating circuits.
Sketch the output waveforms of differentiating and integrating networks for rectangular pulse train inputs.

In the previous year's work we have considered the results of applying sinusoidal voltages to circuits containing resistance, inductance and capacitance and combinations of these. This alternating voltage theory has applications to both low-frequency power and audio circuit systems and to high-frequency radio oscillator and amplifier networks. There are many systems, however, in which the applied voltages are by no means sinusoidal—radar, television and computer systems being familiar examples.

As we have already seen from studies at level 2, when the applied voltage is sinusoidal, the voltages developed across the circuit components are also sinusoidal, differing only in magnitude and phase from the input. The magnitudes and phases depend in turn upon the relative values of resistance or reactance of the components and the frequency of the supply. When the applied voltage is non-sinusoidal the voltages produced across the circuit components are no longer copies of the input waveform but distorted forms of it.

The degree of distortion introduced depends upon the circuit configuration and the relative magnitudes of the circuit component values. This complicates the analysis of the circuit, each form of input voltage calling for a separate and individual investigation.

In this Unit section we consider the results of applying a particular form of non-sinusoidal voltage to simple circuits containing resistance and capacitance only.

SOME NON-SINUSOIDAL WAVEFORMS

There are a number of non-sinusoidal waveforms that are important in the study of electronics. One of these is the *square wave*, illustrated in *Figure 1.1(a)*. We have briefly discussed an oscillator which can generate a waveform of this kind in the previous work at level 2 (see *Electronics for Technicians 2*, section 6) and this will be extended in Section 6 of the present volume. For the waveform itself we define some of the terms associated with it.

For the perfect square wave, the distinctive characteristics are shown in the diagram at (*a*). There is an *instantaneous* rise from a steady low voltage level to a steady high voltage level; the voltage remains at this high level for a given time, then *instantaneously* falls to the low level where it remains for an equivalent length of time before once again rising instantaneously to the high level.

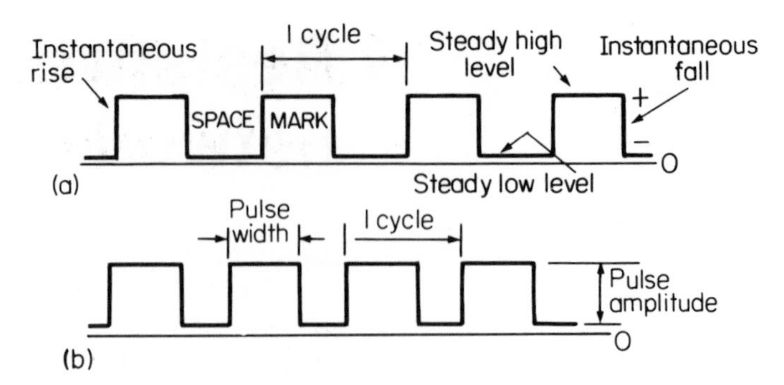

Figure 1.1

The periodic time is, as for the sinusoidal wave, the time taken for one complete cycle of the waveform to occur, and the number of complete cycles occurring in a time of 1 s is the frequency of the wave. Then, in the usual way, we have

$$\text{Frequency} = \frac{1}{\text{period}}$$

Figure 1.1(b) shows the more general *rectangular* waveform of which the square wave is just a particular case. Here the time spent at the high voltage level is greater than the time spent at the low level, though the converse could equally well apply. The ratio of the time that the wave spends at the high level to that at the low level is known as the *mark-space ratio*. The *pulse width* is defined as the time that the wave spends at the high level, and the *pulse amplitude* is the voltage difference between the high and low voltage levels. All these terms are illustrated in the diagram.

Keep in mind that 'high' and 'low' are relative terms; there is no reason why the high level should not, for example, be zero voltage with the low level at, say, −10 V. All the terms we have already defined are unaffected by this.

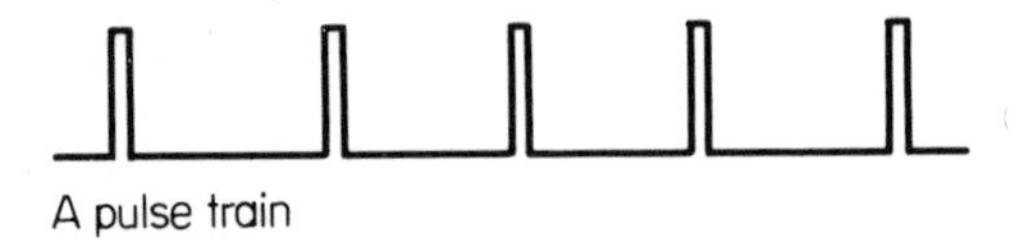

Figure 1.2

In radar and television applications it is quite common for the pulse width to be of extremely short duration, 1 μs or less, with intervening spaces which may be thousands of times as long. It is then customary to speak of the rectangular waveform as a *pulse-train*, and *Figure 1.2* illustrates such a waveform.

(1) Complete the following statements:
(a) A square wave has a mark-space ratio of . . . .
(b) A rectangular wave having a mark-space ratio of 3:1 spends 0.1 s at the high voltage level and . . . . s at the low voltage level.

(c) A rectangular wave switches between voltage levels of −7 V and +3 V. Its pulse amplitude is . . . . V.

(d) The pulse width of a 100 Hz square wave is . . . . s.

Another non-sinusoidal waveform which finds many applications in electronics is shown in *Figure 1.3.* This, as its appearance implies, is a *sawtooth* waveform, characterised by the relatively long period during

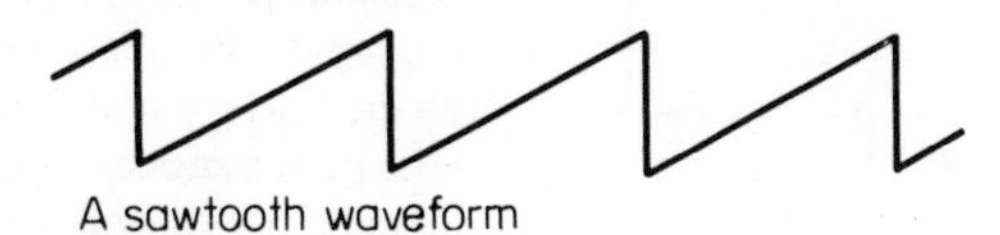

Figure 1.3

which the voltage rises linearly from a low to a high level, followed by an instantaneous return to the low level, after which the cycle repeats. The sloping part of the waveform is known as a *ramp* voltage.

RISE AND DECAY TIME

You might perhaps have already made a mental note that the waveforms described above have been theoretical concepts insofar as we have introduced the idea of an *instantaneous* rise and fall in the voltage levels concerned. Nothing can take place in no time at all, as an event must occupy some finite length of time, however small. If it were otherwise, the event could have no existence for its beginning and its end would be coincident. So with the rise and fall of voltage levels, some time must elapse while the voltage is changing from one level to the other, for if it did not the voltage would be simultaneously at two distinct levels, and this does not accord with commonsense. This fact should be kept in mind when the word 'instantaneously' is used in the following pages.

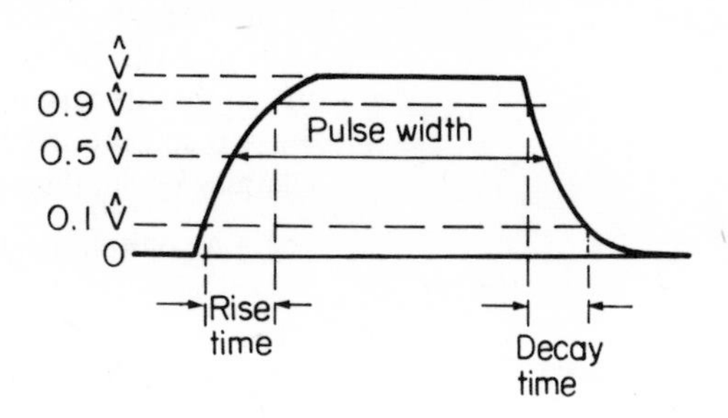

Figure 1.4

Although the rise and fall in voltage has been deliberately exaggerated for clarity, a real square wave when examined very carefully will be found to look rather like that shown in *Figure 1.4.* The rise of voltage is not instantaneous, neither is its decay at the end of the pulse. The *rise-time* is defined, as the diagram illustrates, as the time required for the voltage to increase from 0.1 to 0.9 of its maximum value; the *decay-time* is defined as the time required for the voltage to fall from 0.9 to 0.1 of its maximum value. The pulse width is now defined as being the time interval between those points where the amplitude is 50% of the total amplitude.

THE CHARGE OF A CAPACITOR

Before we can investigate what happens when a rectangular wave is applied to a circuit containing resistance and capacitance, it is necessary to understand the process of changing a capacitor by way of a resistance from a d.c. supply.

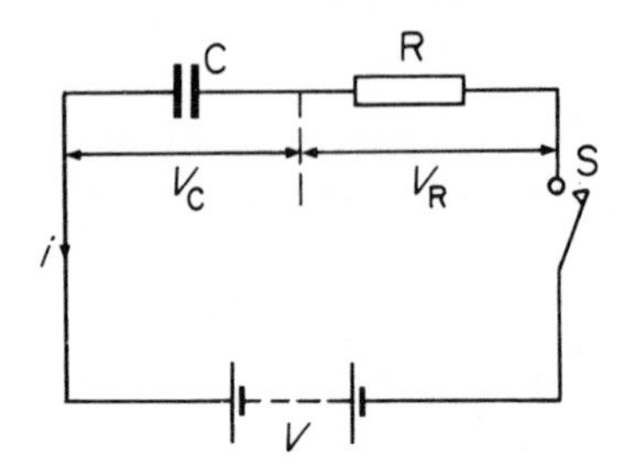

Figure 1.5

In *Figure 1.5* a battery, V volts, is wired in series with a capacitor C and resistor R. When the switch S is closed, the voltage across C does not immediately rise to the voltage level V, since a movement of electric charge is necessary and the expression for charge, $Q = It$, implies that the capacitor needs time to charge. Immediately after switch on, a displacement current begins to flow around the circuit. Electrons move into the plate of the capacitor connected to the negative terminal of the battery and flow out of the plate connected to the positive terminal, a process which leads to the capacitor acquiring charge and hence a rise of voltage across its terminals.

The only opposition to the flow of this current at the *instant* the switch is closed is represented by the resistor R. So the instantaneous value of the current at the moment of switch-on is simply that given by Ohm's Law, V/R. The capacitor has no charge and hence no voltage across its terminals; and since the sum of V_C and V_R must at all times equal V, and V_C is zero, all the voltage must instantaneously appear across R. We will indicate this initial value of the current by I_O.

This initial current does not continue unchanged as it would do if the capacitor was not there. C begins to charge immediately and the voltage across its terminals rises accordingly. The voltage across R correspondingly falls by an equal amount, so that for a given V_C $V_R = V - V_C$. Consequently, at any particular time after switch-on, the charging current will be *less* than I_O and will be given by

$$i = \frac{V - V_C}{R}$$

So, as the voltage across C rises, both the current and the *rate* of charging fall. Hence C charges progressively more and more slowly as time passes.

Now with the passage of time, the capacitor must eventually 'fill up' so that its terminal voltage V_C becomes equal to V. In this situation $V_R = 0$ and the circuit current is reduced to zero. Throughout the entire charging cycle, therefore, we can make the following observations:

At switch-on ($t = 0$), $V_C = 0$, $V_R = $ V and $I_O = \frac{V}{R}$

At any time during the charge: $V_C > 0$, $V_R = V - V_C$ and

$$i = \frac{V - V_C}{R}$$

At the completion of the charge: $V_C = V$, $V_R = 0$ and $i = 0$.

So the capacitor voltage has risen from zero to V volts and the resistor voltage and circuit current have both fallen from V_R and I_O respectively to zero.

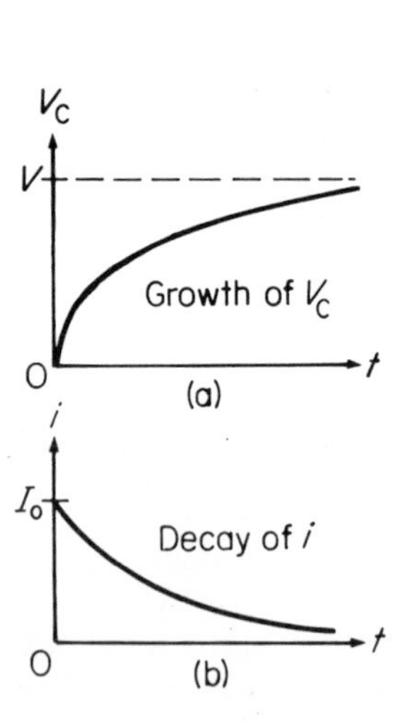

The way in which the capacitor voltage rises with time is shown in *Figure 1.6(a)* with the fall of circuit current illustrated at (b). Both of these curves have the same mathematical form and in the Electrical Principles course you will be shown how a mathematic expression

Figure 1.6

can be derived for them. For our present investigation we need know them only as *exponential* growth (V_C) and decay (i) curves and keep their general appearance in mind.

Here are some worked and self-assessment problems to help fix these ideas in your mind.

Example (2) A capacitor is connected in series with a 10 kΩ resistor across a 200 V supply. What is the initial current and the voltage across the capacitor when the circuit current is 5 mA?

$$\text{At switch-on the initial current } I_o = \frac{V}{R} = \frac{200}{10000} \text{ A}$$

$$= 20 \text{ mA}$$

This current decays towards zero as the capacitor charges; when it has fallen to 5 mA, the voltage across R at that instant is

$$V_R = iR = 5 \times 10^{-3} \times 10000 = 50 \text{ V}$$

Hence

$$V_C = V - V_R = 200 - 50 = 150 \text{ V}$$

(3) When a capacitor and series resistor are connected to a 50 V supply, the initial current is 100 mA. Calculate the value of the resistor and the circuit current at the instant when the voltage across C is 20 V.

(4) A voltmeter whose resistance is 20 kΩ is connected in series with a capacitor across a 100 V supply. What will the voltmeter read (a) at the instant of switch-on; (b) at the instant the circuit current is 2 mA; (c) when the capacitor is fully charged?

THE TIME CONSTANT

As shown in *Figures 1.6(a) and 1.7* the rate of rise of voltage across C becomes progressively less as the charge proceeds. The gradient of the curve at any point is defined as the slope of the tangent drawn to the curve at the point in question or, in mathematical form, as $\delta V_C/\delta t$, where δV_C is a small increment in capacitor voltage over a small increment in time δt.

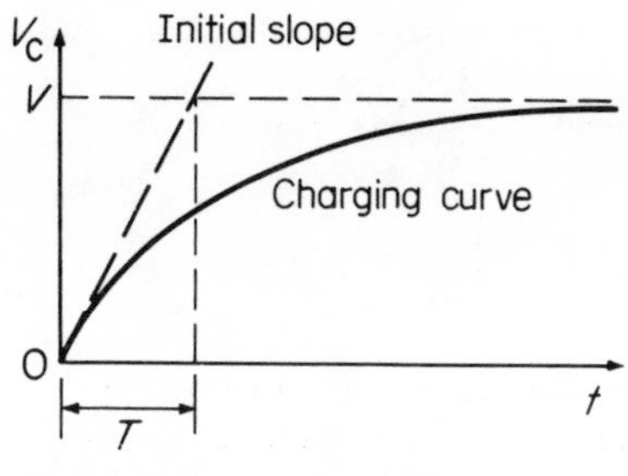

Figure 1.7

Now $Q = CV = it$ so that a small charge δQ due to current i flowing for a time δt produces a small change of voltage δV_C. So

$$\delta Q = C.\delta V_C = i.\delta t$$

and

$$i = C.\frac{\delta V_C}{\delta t}$$

But $\delta V_C/\delta t$ is the gradient of the charging curve (or the rate of change of voltage). When the switch is closed, $t = 0$ and the initial current

$$I_O = \frac{V}{R} = C.\frac{\delta V_C}{\delta t_O}$$

$$\text{So the } \textit{initial} \text{ gradient } \frac{\delta V_C}{\delta t} = \frac{V}{CR} \text{ V/s}$$

and this is shown as the dotted line rising from the origin of *Figure 1.7*. If the charge was to *continue at this rate* time taken for V_C to reach the final level V would be T seconds.

But

$$T = \frac{\text{voltage}}{\text{rate of increase of voltage}} = \frac{V}{V/CR}$$

$$\therefore \quad T = CR \text{ seconds}$$

This product is known as the *Time-Constant* of the circuit, and it is a very important concept.

C and R must be expressed in the proper units, farads and ohms, for the time-constant to be in seconds. However, if C is expressed in μF, the time will be given in μs. It is often easiest to express C in μF and R in MΩ, and the answer is then once again in seconds.

Example (5) What is the time-constant of a 3 μF capacitor in series with a 100 kΩ resistor?
Working in Farads and ohms we have

$$T = 3 \times 10^{-6} \times 100 \times 10^3 = 0.3 \text{ s}$$

Alternately, working in μF and MΩ

$$T = 3 \times 0.1 = 0.3 \text{ s}$$

Example (6) A 10 μF capacitor and a 2 MΩ resistor are connected to a 100 V supply. Find (a) the time-constant; (b) the initial current; (c) the initial rate of rise of voltage across C.

(a) $T = CR = 10\ \mu\text{F} \times 2\ \text{M}\Omega = 20\ \text{s}$

(b) $I_O = \frac{V}{R} = \frac{100}{2 \times 10^6}\ \text{A} \quad = 0.05\ \text{mA}$

(c) Inital rate of rise of voltage $= \frac{V}{CR}$ V/s

$$= \frac{100}{20} = 5 \text{ V/s}$$

(7) A circuit consisting of a 50 μF capacitor in series with a resistor R is to have a time-constant of 2 s. What value of resistor is needed? If the circuit is connected to a 100 V supply, find (a) the initial current; (b) the rate at which the voltage is rising when the capacitor is charged to 50 V.

(8) Sketch curves of capacitor voltage against time for circuits having (a) very long time-constants, (b) very short time-constants.

SKETCHING THE CHARGING CURVE

We can use our knowledge of the time-constant $T = CR$ to obtain an approximate graph of capacitor voltage against time without having to resort to more advanced mathematics.

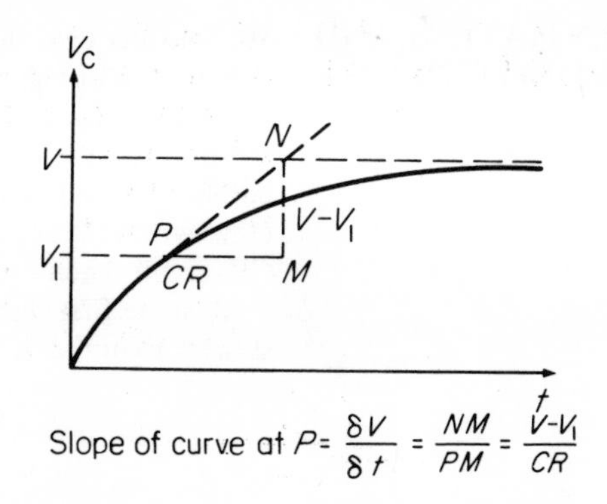

Figure 1.8

At any point on the curve, such as P in *Figure 1.8*, when the voltage has risen to V_1 volts

$$\text{charging current } i = \frac{V - V_1}{R} = C.\frac{\delta V}{\delta t}$$

and the gradient of the curve, represented by the tangent PN is

$$\frac{\delta V}{\delta t} = \frac{V - V_1}{CR}$$

Hence, in the triangle PNM, the side $NM = V - V_1$ and the side $PM = CR$, the time-constant. Clearly, this argument will be true for *any position* on the curve, hence for any point on the curve, if the charge then continued at that instantaneous rate, the time in which the charge would be completed would be CR s, the time-constant. You should now be able to deduce that in theory the capacitor can never be completely charged! Also, you should be able to work out for yourself how you can obtain an approximate graph of the charging curve.

THE DISCHARGE CURVE

When a capacitor C, charged to a voltage V, is connected to a resistor or R there is an initial current of discharge given by $I_o = V/R$. The voltage across the capacitor falls at its maximum rate, see *Figure 1.9*, and would reach zero in a time $T = CR$ s if the fall continued at that

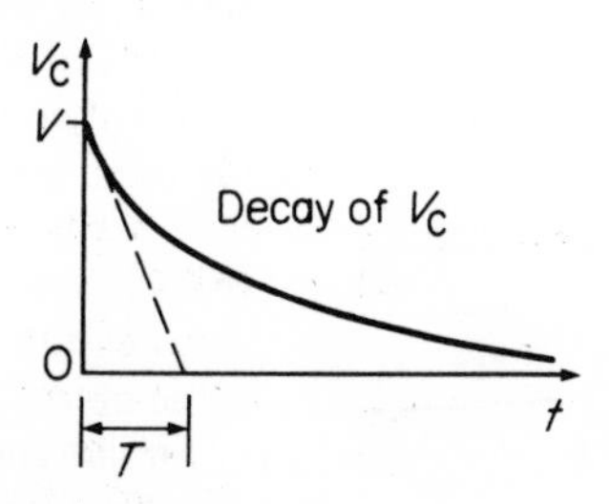

Figure 1.9

As the voltage falls, however, the current through R and the rate of change of voltage, $\delta V/\delta t$, both fall also. The curve, which is applicable to either the discharge current or the voltage across C, is of similar form to the charging curve and has similar time-constant properties.

DIFFERENTIATING AND INTEGRATING CIRCUITS

We consider now the application of a voltage pulse whose time variation is as shown in *Figure 1.10* to a series *CR* circuit. It is convenient to break the effects of the pulse into two parts: firstly the effect of the abrupt increase in voltage at the start of the pulse (the switch-on condition), and secondly the effect of the abrupt decrease at the end, (the switch-off condition). We have already noted what happens in the *CR* circuit when these two cases are considered separately.

Assume first that the time-constant of the circuit is very much shorter than the duration *P* of the pulse. Then the charging of the capacitor is completed well before the pulse ends, and the capacitor voltage remains at the peak level of the input V_i for the remainder of the pulse time *P*, see diagram *Figure 1.10(b)*. The corresponding variation in the voltage across *R* is shown at (c); V_R rises instantly to the input pulse level and thereafter rapidly decays away as V_C rises, remaining at the zero level for the remainder of the pulse time *P*.

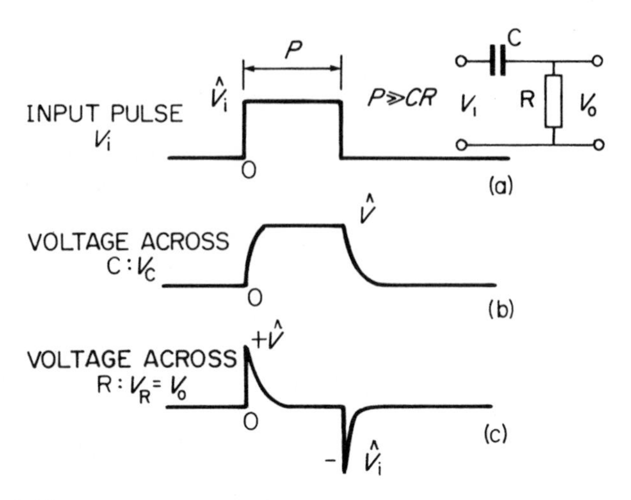

Figure 1.10

When the input pulse falls to zero, the capacitor must discharge through *R* since the input terminals are now effectively short-circuited. This it does in a very short time as shown at (b); the voltage across R instantaneously follows the fall in input (which is now effectively in a negative direction), and then decays from this negative value back to zero as shown at (c).

The output voltage developed across *R* consists of two sharp narrow pulses of voltage of opposite sign, one at the start and one at the end of the input pulse. The *duration* of these narrow pulses depends upon the *CR* time-constant, but *not* upon the duration (pulse width) of the applied pulse. This pulse width determines only the time which elapses between the leading edges of the first and second output pulses developed across *R*.

Suppose now that the circuit time-constant is very much greater than the pulse duration *P*. In this case the voltage across *C* will rise very slowly and will have reached an amplitude that is only a fraction of the input amplitude $\hat{V}_i$ before the input pulse drops to zero and causes the capacitor to discharge. *Figure 1.11(a) and (b)* shows this new situation. For example, let us assume that the capacitor has charged only to one-fifth the input amplitude when the input falls to zero, that is, to a level

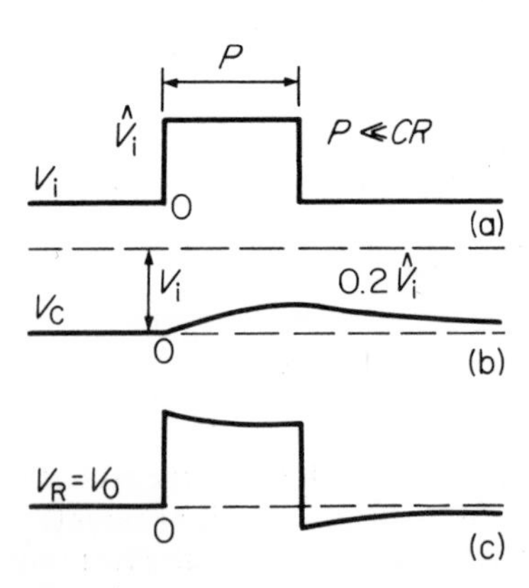

Figure 1.11

$0.2\hat{V}_i$. After time P the capacitor discharges from this level towards zero, but the discharge rate is quite different from the preceding charge rate. The rate of charge or discharge is proportional to the voltage across the resistor.

At the commencement of the charging period this voltage is (instantaneously) $\hat{V}_i$ but at the beginning of the discharge period it is only (in our example) one-fifth or $0.2\hat{V}_i$. Thus the rate of discharge is only one-fifth the rate of charge and the discharge curve exhibits a very long tail.

The voltage across R rises to a value $\hat{V}_i$ at the onset of the rectangular input pulse, then falls at the same rate as the capacitor voltage rises (very slowly remember) for the duration of the pulse, and then drops instantaneously by an amount *equal* to $\hat{V}_i$ at the end of it. Finally, the resistor voltage rises exponentially towards zero at the same rate as the capacitor voltage falls. As *Figure 1.11(c)* shows, the shape of the output pulse developed across R is a close replica of the input pulse. The *longer* the circuit time-constant relative to pulse duration P the *more closely does the output correspond to the input.* The voltage across C, on the other hand, produces a low amplitude (and distorted) pulse many times the length of the input pulse.

Summarising the above gives us the following conclusions:

(i) The maximum value of the output voltage developed across R is in both cases equal to he amplitude $\hat{V}_1$ of the applied pulse.

(ii) The shape of the output voltage is a close approximation to the input if $CR \gg P$, but if $CR \ll P$ the input pulse is converted into two narrow pulses of opposite sign and whose durations depend upon the value of the time-constant.

The terms 'differentiating' and 'integrating' are given to the circuits in which $CR \ll P$ and $CR \gg P$ respectively. For differentiation the time-constant must be short and the output is taken across the resistor. The narrow spikes developed across R are then approximately proportional to the time rate-of-change of the input voltage. For integration the time-constant must be long and the output is taken across the capacitor. The broad pulse developed across C is then approximately proportional to the time-integral of the input voltage.

Example (9) What would the true differentiated version of a rectangular voltage pulse look like, and how closely does a practical differentiating circuit approximate to it?

For the rectangular pulse shown in *Figure 1.12*, the instantaneous rate of change (or gradient) dV_i/dt is either infinite or zero. It is infinite on the leading and trailing edges of the pulse where the waveform gradient is vertical, and zero elsewhere where the waveform is horizontal and the gradient zero. A true picture of the differentiated version of the rectangular pulse consequently is as shown in *Figure 1.12(b)*.

The output of a practical differentiating circuit is shown at *Figure 1.10(c)* and although this consists of narrow spikes of voltage, the similarity with the true differentiated voltage is clear. For an extremely short time-constant the similarity becomes very marked and the term 'differentiating circuit' is fully justified.

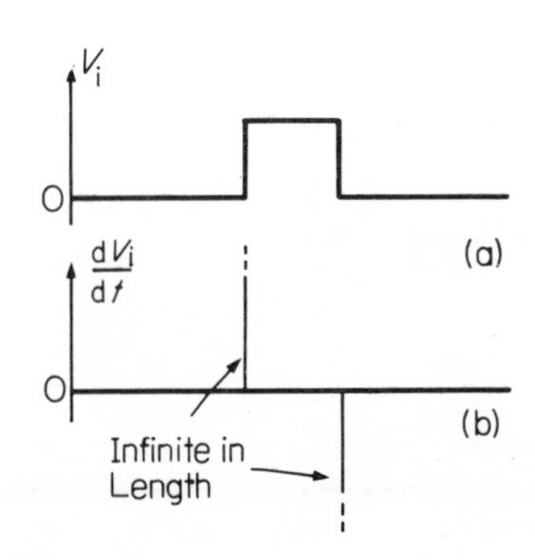

Figure 1.12

APPLICATION OF PULSE TRAINS

We now look into the case where a succession of rectangular pulses is applied to differentiating and integrating circuits in turn. Suppose P_1 is the duration of each pulse and P_2 the interval (space) between the pulses. The result of applying such a pulse train to a differentiating circuit where $CR \ll P_1$ is shown in *Figure 1.13*. The shapes of the voltages across C and across R are similar to those already described, and the output across R is a succession of alternate positive and negative spikes. There is nothing much to add to what has already been said about this kind of circuit.

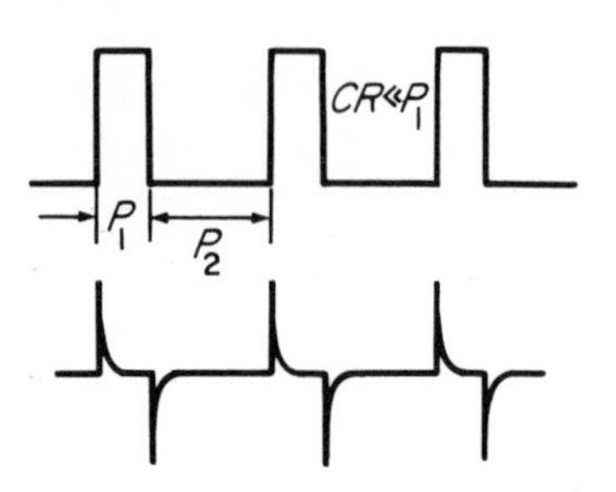

Figure 1.13

Turning now to the integrating circuit, some new points for discussion emerge. Let the pulse train be applied to the integrating circuit and the voltage across C be examined.

Again, the general shapes of the voltages developed across C and across R are similar to those already described, but this time the long discharge tail of the capacitor is not permitted to fall away to zero.

As *Figure 1.14(b)* shows, the mean level of voltage across C tends to rise with each successive pulse because the capacitor has barely started to discharge before the following pulse arrives. After a number of pulses have occurred, a steady state is reached in which the charge acquired by the capacitor during the pulse time P_1 is exactly *equal* to the charge lost during the interval P_2. This means that the mean current flowing into C through R is zero when equilibrium has been reached, so that the mean value of V is also zero.

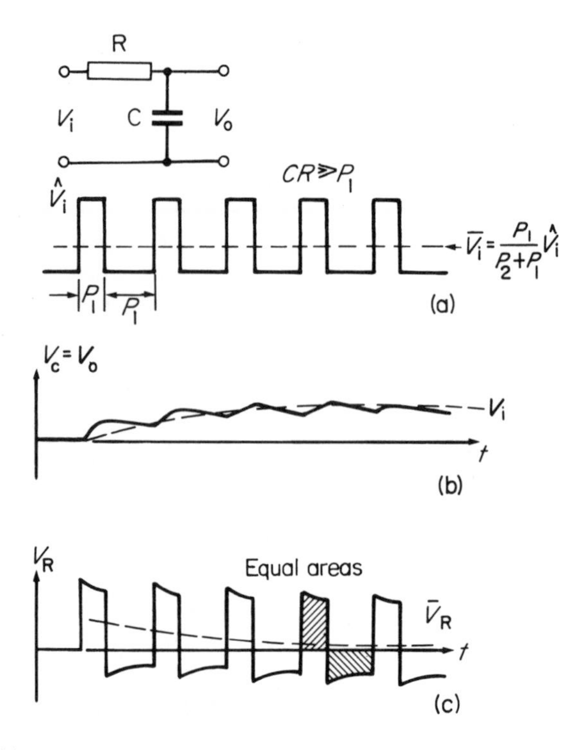

Figure 1.14

The graph at (c) shows the voltage developed across R and at equilibrium the area (representing voltage × time) enclosed above the mean (zero) voltage line during the time P_1, is equal to the area enclosed below this line during the time P_2.

The mean output across C, shows as a dotted line in diagram (*b*), is seen to be a long, single pulse: the circuit has integrated (or summed up) all the separate input pulses. The voltage output across R is a good replica of the input voltage train, but the waveform is balanced about a zero level unlike the input which is positioned wholly above the zero level. This effect comes about because of the removal of the d.c. component, or the reference level if you wish.

The succession of positive input pulses can be considered to be made up of a steady (d.c) component of voltage and a purely alternating component of voltage. The mean value $\bar{V}_i$ of the input voltage is clearly

$$\bar{V}_i = \frac{P_1}{P_1 + P_2}\hat{V}_i$$

The alternating component has a mean value of zero. When a steady voltage is applied to a series CR circuit, the whole of the voltage appears, as we have previously seen, ultimately across C and the rise is exponential. In the present case therefore, the mean voltage across C increases exponentially with time-constant CR towards the value

$$\frac{P_1}{P_1 + P_2}\hat{V}_i$$

and ultimately the mean voltage across R becomes zero.

It is clear then that if $CR \gg P_1$ and $P_1 < P_2$ the ultimate output across R is a replica of the input voltage insofar as the alternating component of the input is concerned; but that the steady (d.c) component of the input is not reproduced.

In many circuit applications it is often necessary to re-insert the steady component at the output, that is, the output pulses across R have to be positioned on a zero base line as they were originally in the input. The process of doing this is known as *d.c. restoration.*

Example (10) Suppose P_2 to be less than P_1 for the input pulse train. What effect would this have on the output waveforms?

If $P_2 < P_1$ the input can be considered as a succession of *negative-going* pulses and an argument similar to that used above can be employed. The condition for a replica of the alternating component of the applied pulses to appear across R is that $CR \gg P_2$.

Example (11) The input to the circuit of *Figure 1.15(a)* is the rectangular waveform of *Figure 1.15(b)*. Sketch the output waveform, marking the high and low voltage levels attained and indicating the new zero level.

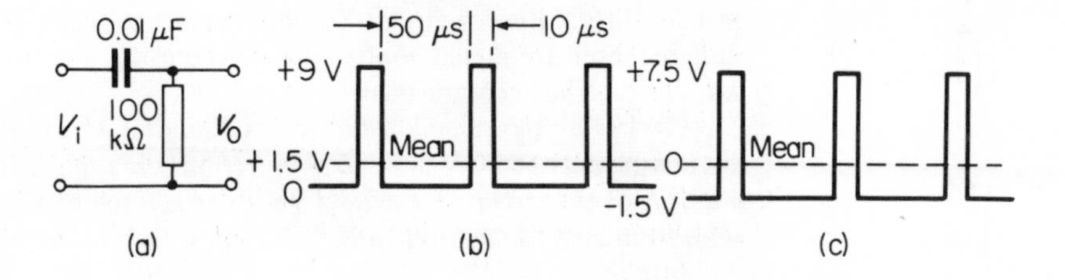

Figure 1.15

In the circuit shown, the time constant $CR = 0.01 \times 0.1 = 0.001$ s or 1000 μs. As CR is long compared with the period between the pulses, 50 μs, the shape of the waveform will be practically unaffected. The average value of the input wave is

$$\overline{V_i} = \frac{P_1}{P_1 + P_2} \cdot \hat{V}_i = \frac{10}{60} \times 9 = 1.5 \text{ V}$$

The output wave must 'balance' itself so that its average is zero, therefore the wave will vary between −1.5 V and +7.5 V as shown in *Figure 1.15(c)*.

Your first short selection of problems based on the work of this unit section now follows.

PROBLEMS FOR SECTION 1

(12) Complete the following statements:

(a) A rectangular wave spends 100 ms at a high voltage level and 220 ms at a low voltage level. The period of one cycle is . . . . and the mark-space ratio is . . . .

(b) A rectangular wave switches between voltage levels of −10 V and +3 V. Its amplitude is . . . . and its mean value is . . . .

(c) The pulse width of a 10 kHz square wave is . . . . and its period is . . . .

(13) A rectangular pulse rises from 0 V level to +5 V level in 1 μs. After 100 μs the level returns to 0 V in 2 μs. Find (a) the rise time, (b) decay time; (c) pulse amplitude; (d) pulse duration (width).

(14) To a suitable time axis, sketch a pulse waveform having the following characteristics: rise time 2 μs, decay time 3 μs, pulse width 10 μs and periodic time 20 μs.

(15) A constant voltage of 100 V is maintained across a series circuit of 0.5 μF and 25 kΩ. Sketch the variation with time of the voltage across C and R. From your sketch (or by calculation) find the rise and decay times of the voltages.

(16) Why are differentiating and integrating circuits so called?

(17) A 4 μF capacitor and a 2 MΩ resistor are joined in series across a 100 V d.c. supply. Find (a) the initial charging current, (b) the rate of rise of voltage across C at switch on, (c) the rate of rise of voltage at the instant when the capacitor is charged to 50 V.

(18) A rectangular pulse is applied to an integrating circuit. Sketch the output waveform when (a) the pulse width is greater than the circuit time-constant, (b) the pulse width is less than the circuit time-constant.

(19) If a 0.1 μF capacitor and a 100 kΩ resistor are to be used as (a) a differentiating circuit, (b) an integrating circuit, suggest suitable input frequency limits in each case if the function of the circuits is to be accomplished.

(20) Define the time-constant of a CR circuit. If a square wave oscillating between 0 V and −6 V at 500 kHz is applied to the input A of the circuit of *Figure 1.16*, draw a graph of the voltage waveform at B for the following values of C and R: (a) 0.001 μF

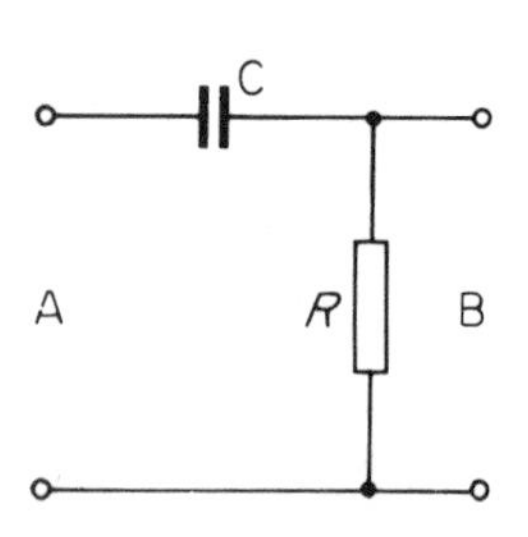

Figure 1.16

and 500 Ω, (b) 0.001 μF and 1 kΩ, (c) 0.0001 μF and 2 kΩ, (d) 0.001 μF and 1 MΩ (C. & G.)

(21) A telephone dial may operate between 7-12 pulses per second with a break percentage of 63-72%. What is the duration of (i) the minimum pulse break; (ii) the minimum make pulse? (C. & G.)

2 Small-signal voltage amplifiers

Aims: At the end of this Unit section you should be able to:
Define the main characteristics of amplifiers.
List the biasing conditions for Class A, B and C operation and state their applications.
Describe the three circuit configurations and estimate the gain of a Class A common-emitter amplifier from the transistor parameters and by graphical analysis.
Describe and explain the operation of resistance-capacitance, transformer and directly coupled amplifiers.
Sketch and use simple equivalent circuits.
List the basic amplifier measurements and describe simple measuring circuits.

AMPLIFIER CHARACTERISTICS

Although in this section and also section 3 we shall be interested in voltage amplifiers operating under Class A conditions, we begin with a general survey of amplifier characteristics and the terminology used. Some of this may appear confusing at first reading, but it is important that the contents of this introductory section are understood and appreciated before an analysis of a specific type of amplifier is attempted.

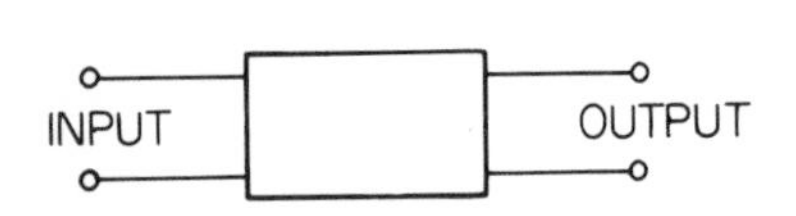

Figure 2.1

In very general terms, an amplifier is a device where a small input signal is made to control a larger output signal. We can represent an amplifier as a box having two input and two output terminals as shown in *Figure 2.1*. The box will contain some active device or devices, transistors or valves, together with some associated passive components such as resistors and capacitors, and a source of d.c. power derived from batteries or rectified and smoothed a.c. mains. There will also be a *load resistance* across which the output signal voltage will be developed. The conventional amplifier therefore, may be represented in the form shown in *Figure 2.2(a)*; the source of d.c. power is connected by way of the load resistor to the active control device. The load does not have to be a resistor as such, it may take the form of a tuned circuit, for example, or perhaps an inductor.

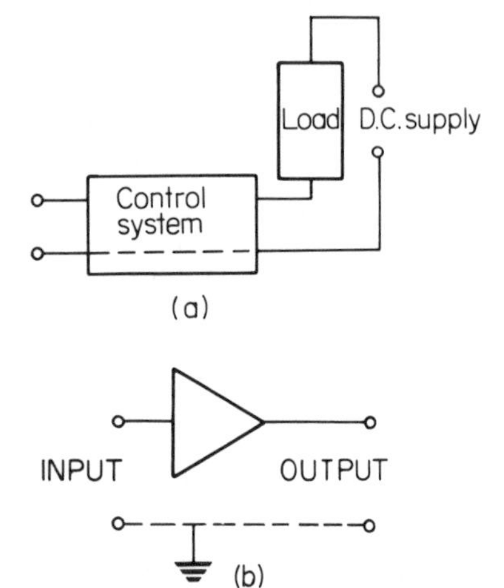

Figure 2.2

One of the pair of input terminals of an amplifier is usually joined to a common or earth line to which one side of the d.c. power supply is also wired. Such an amplifier is said to have a *single-ended* input connection. The same also applies to the output terminals, one of these terminals may be effectively at earth potential and the output is then said to be single-ended.

The conventional symbol for an amplifier is shown in *Figure 2.2(b)* where the arrow is shown pointing in the direction of signal flow. The diagram illustrates a single-ended input and output amplifier; the dotted common earth connection is often omitted in circuit diagrams using these symbols. This can make some diagrams difficult to read for anyone without a lot of experience, so we shall retain the diagram for an amplifier as a rectangular box with both input and output pairs of terminals clearly shown.

TYPES OF AMPLIFIER

Amplifiers can be classified into two main types:

1. *Small-signal amplifiers* which are designed to amplify small input signals having voltage levels of the order of a few microvolts to a few millivolts. An amplifier which immediately follows a record pick-up head or a microphone would be an example of a small-signal amplifier, as would an amplifier whose input was the very small radio-frequency signals received on an aerial. It is easier to avoid distortion in amplifiers of this kind than it is in the second category below.
2. *Power amplifiers* are designed to operate with very large input signal voltages, at least of the order of several volts. Their output requirements are large voltage and current variations so that considerable *power* is available for driving such devices as loudspeakers or, in industrial applications for example, electric motors. It is difficult to obtain large power outputs of this sort without the introduction of unwanted distortion.

A single amplifier stage containing one transistor or one valve is rarely sufficient to provide the overall amplification needed. Two or more devices are then connected in *cascade* (or tandem), the output of one being fed as input to the next, and so on. The signal is then progressively amplified as it passes through the system. At the end, when a sufficient amplification in the small-signal category has been obtained, a power amplifier is introduced to provide the final output level required. We shall deal for the time being with small-signal amplifiers only.

BASIC REQUIREMENTS

We expect an amplifier to fulfil three basic conditions and we will deal with these in turn in the following paragraphs.

1. Amplifier gain

The output signal will be greater in voltage or current amplitude than the corresponding input signal quantity*. Amplifier gain is a measure of the voltage or current amplification provided by the amplifier. We define

$$\text{Current gain } A_i = \frac{\text{the change in output signal current}}{\text{the change in input signal current}} = \frac{\delta i_o}{\delta i_i}$$

$$\text{Voltage gain } A_v = \frac{\text{the change in output signal voltage}}{\text{the change in input signal voltage}} = \frac{\delta V_o}{\delta V_i}$$

Both A_i and A_v are ratios of like quantities and are therefore pure numbers. Signal currents and voltages are normally expressed in r.m.s. values but in many cases it is just as convenient (and sometimes necessary) to work in peak values. For sinusoidal waves the resulting figures obtained for A_i and A_v are identical.

*It is perfectly possible for an amplifier to be designed to give a smaller signal output than its input, and it is still referred to as an amplifier. We shall assume throughout our notes that the output signal amplitude is greater than the input signal amplitude.

Example (1) The charge in the output voltage of an amplifier is 3 V r.m.s. when its input changes by 15 mV r.m.s. What is the voltage gain?

$$A_v = \frac{\delta v_o}{\delta v_i} = \frac{3v}{0.015v} = 2000$$

It is often necessary to know the power gain of an amplifier. Let the output power $P_0 = V_0 i_0$ watts for an input power $P_i = V_i i_i$ watts. Then the power gain is given by the ratio of output to input power, or

$$A_p = \frac{P_o}{P_i} = \frac{V_o}{V_i}\frac{i_o}{i_i} = A_v.A_i$$

So power gain is simply the product of voltage and current gain. Power gain is always taken as a positive quantity, whereas voltage or current gains are sometimes given a negative sign to indicate phase conditions.

The total voltage gain of a complete amplifier is the product of the gains of its individual transistor or valve stages. If, for example, three transistors are connected in cascade and have individual voltage gains of 10, 20 and 5, the total overall gain of the system is $10 \times 20 \times 5 = 1000$. It is not possible to obtain the overall *current* gain of an amplifier in this way because the output current of any particular stage is not applied as input to the next. If often happens that the figures involved in gain calculations become large and clumsy, for example, when an amplifier shows large gain variations with input signal frequency and when these changes have to be plotted as a graph.

It is easy to imagine the problem of showing a range of gain ratios between say, 10 and 1 000 000 on an ordinary piece of graph paper. For this reason the gain ratios are expressed in logarithmic units or decibels, and you will recall from your work in the electrical principles course that

$$\text{Voltage gain} = 20 \log \frac{v_o}{v_i} \text{ dB}$$

$$\text{Current gain} = 20 \log \frac{i_o}{i_i} \text{ dB}$$

$$\text{Power gain} = 10 \log \frac{P_o}{P_i} \text{ dB}$$

By using decibel gain units very large variations in the gain ratios are brought down to manageable size. The gains can also be directly added (or subtracted) instead of being multiplied (or divided) when a large number of stage gains are involved. The following table shows how effective this compression of large numbers can be:

Voltage or current gain ratio	Gain in dB
10	20
100	40
1000	60
10000	80
100000	100
1000000	120

Example (2) Express as decibel gains, voltage ratios of (a) 500, (b) 75, (c) 2.

(a) Gain = 20 log 500
= 20 × 2.7 = 54 dB

(b) Gain = 20 log 75
= 20 × 1.875 = 37.5 dB

(c) Gain = 20 log 2
= 20 × 0.301 = 6 dB

The corresponding power gains for the given ratios in terms of power would each be one-half of the above figures. It is useful to bear in mind that

6 dB voltage or current gain = ratio 2
3 dB power gain = ratio 2

So if the *voltage* gain of a system falls by 6 dB, or its *power* gain falls by 3 dB, its output voltage or its output power falls to *one-half* of its original value.

(3) Show that a 3 dB voltage reduction is equivalent to a fall in the voltage to 0.707 of its original value.

2. Bandwidth Amplifiers are usually classified by stating the range of signal frequency over which they are designed to provide a specified gain. Some amplifiers are made to amplify at some fixed, single frequency; others are designed to cover a range of frequencies, very often a considerable range.

A common everyday example is the audio-frequency amplifier. As its name implies, it is designed to amplify signals in the range of frequencies that can be detected by ear, though a good quality amplifier of this sort would amplify signals over a rather wider range than most ears cover—probably from some 10 Hz up to, perhaps, 20 kHz or more. Provided its distortion was very small, this would be called a 'hi-fi' amplifier. A less sophisticated design may only have a range extending only from some 100 Hz to perhaps 10 kHz; this sort of amplifier might be found in portable record players or good quality portable radio receivers.

For speech purposes on telephone systems, amplifiers need only cater for an upper frequency limit of about 3.5 kHz. *Wideband* or *video* amplifiers as they are called are designed for much greater frequency ranges, extending very often to an upper frequency limit of many megahertz with a lower limit which may well be d.c. This type of amplifier is found in television receivers, good quality oscilloscopes and radar systems.

Tuned amplifiers, on the other hand, amplify over only a limited range of frequencies and are basically designed as single frequency amplifiers. These are known as *selective* or narrow-band amplifiers and are found, for example, in the radio-frequency and intermediate-frequency stages of radio receivers. Like gain, the choice of frequency range is all a matter of what the amplifer has to do.

The frequency range of amplifiers can be exhibited on what is called a *frequency response* curve. A typical response curve for an audio-amplifier is shown in *Figure 2.3(a)*. This is a plot of amplifier gain versus frequency. Notice that the vertical axis is scaled in dB units of gain and that the frequency axis is divided logarithmically; these scalings ensure that the necessary wide variations in both gain and frequency can be accommodated on the paper. Notice also that the gain falls off at both the low and high frequency sections of the scale but remains substantially constant over the middle, or mid-frequency, portion of the range. We shall discover why this happens in due course.

In the meanwhile we can define another amplifier characteristic from a consideration of this graph, the *bandwidth*. The bandwidth of an amplifier is defined as that range of frequencies over which the gain has not fallen by more than 3 dB from its mid-frequency gain figure. This

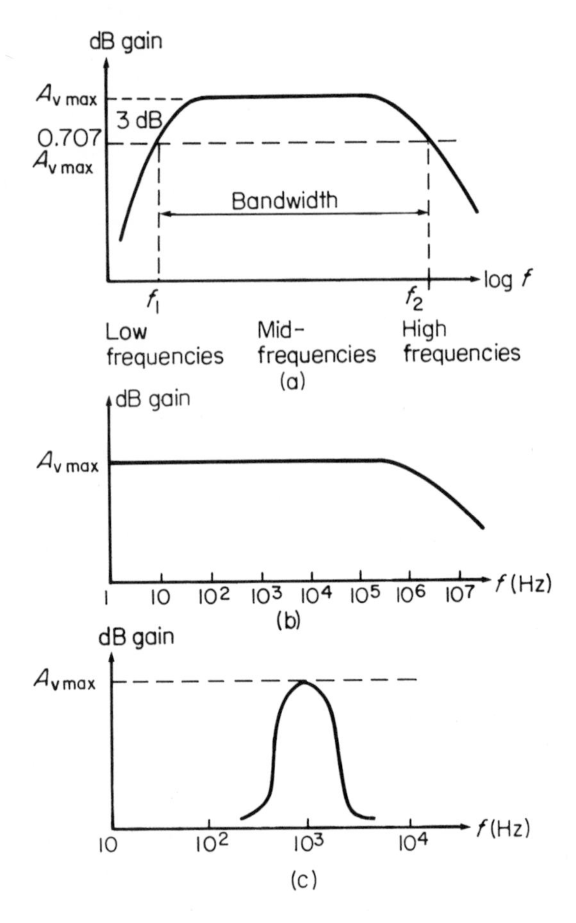

Figure 2.3

may seem an arbitary kind of measurement but the reason for it will also become apparent later on. If, as *Figure 2.3(a)* indicates, the mid-frequency gain is A_{vmax}, the 3 dB points will correspond to 50% of the maximum gain in terms of power, or (if you have worked *Example (3)* above properly) 70.7% of the maximum gain in terms of voltage. From the diagram the bandwidth of the amplifier is given by $f_2 - f_1$ and the level of measurement is $0.707A_{vmax}$. This range is also known as the *passband* of the amplifier.

(4) What kind of amplifiers do the response curves shown in *Figure 2.3(b) and (c)* represent? Make a rough estimate of the bandwidth of each of these amplifiers.

3. Distortion

We should expect an amplifier to provide us with an output signal which is an exact replica, as regards waveform, of the input signal. This means that the signal should not suffer re-shaping or *distortion* during the process of amplification. If a sine wave is applied as input to an amplifier and a square wave comes out, that is gross distortion!

Few amplifiers would be this bad but it is not a simple matter, nevertheless, to design an amplifier having negligible distortion, especially if large power output is demanded, without sophisticated and expensive circuitry. Nor, in many cases, is such extreme purity necessary. An audio-amplifier, for example, can exhibit quite a lot of distortion (of a particular kind) before the output becomes objectionable to the ear.

Some people would even argue that a lot of modern music appears to be sufficiently distorted before it gets anywhere near an amplifier. Be that as it may, distortion in amplifiers does turn up in several different forms and for several different reasons. We shall briefly mention three of the most important:

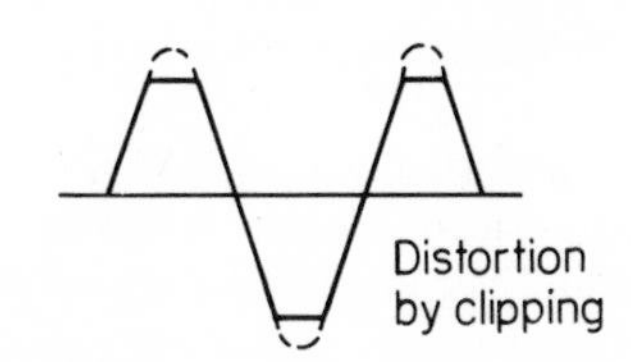

Figure 2.4

(i) *Amplitude Distortion* occurs when the input to the amplifier exceeds a certain magnitude. Clearly, the maximum amount of power that an amplifier can deliver to its load is not unlimited and any amplifier has a fairly well defined maximum output capability. If an attempt is made to obtain an output in excess of this limit by increasing the input signal amplitude, the amplifier becomes *overloaded.* A sinusoidal input wave, for example, would then become flattened at the tops and bottoms as shown in *Figure 2.4.*

This kind of distortion is also known as *non-linear* distortion or *harmonic* distortion because the output waveform is no longer sinusoidal but is composed of a number of sinewave harmonic components of varying phases and amplitudes.

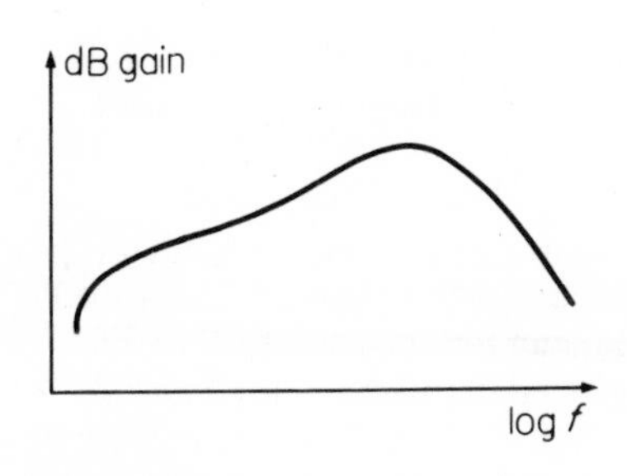

Figure 2.5

(ii) *Frequency Distortion* occurs when the gain of the amplifier is not constant with frequency within the working range. In *Figure 2.3(a)* we noted a gain-frequency response curve that was flat over the passband. An amplifier suffering from frequency distortion may have a response curve of the shape shown in *Figure 2.5.* Here there is a loss of gain at low frequencies and a considerable increase in gain at high frequencies.

The opposite situation might also exist and all sorts of variations in the shape of the curve are possible. Clearly the output signal will be distorted because the many harmonic components of differing frequencies of which it is made up will not be amplified in their proper relative proportions.

(iii) *Phase Distortion.* As well as possessing gain, any amplifier also introduces a phase shift between its input and output terminals. We know, for example, that a transistor in the common-emitter connection inverts the signal voltage input, that is, it introduces a 180° phase shift. It might appear at first that this phase shift would be the same at all frequencies but this is not so. Not only does the phase shift in the transistor itself change as the frequency becomes very high, but the effect of the associated components comes into the picture at frequencies beyond the extremes of the passband and the overall phase shift does not remain at 180°. Hence, as the signal frequency is increased (or reduced) so the phase of the output signal changes relative to the input.

When the input wave is complex, the relative phases of its harmonic components are shifted by varying amounts and the output complex wave can well be totally different in shape from that at the input.

It would be of value to revise (or at least read up) the relevant work on complex waves in your Electrical Principles course 3 at this stage. (See section 9 of *Electrical and Electronic Principles 3*).

CLASSES OF OPERATION

Classes of operation is yet another category into which amplifiers may be sub-divided. There are three basic classes:

Class A in which the transistor or valve is biased so that a mean or *quiescent* current flows in it all the time. The transistor or valve current is varied by the signal about this mean value so the device is conducting throughout the whole of the input cycle. This class of operation is invariably used in small-signal voltage amplifiers and very often in low-power audio amplifiers.

Class B. Here the transistor or valve is biased to the point at which the current through it is just cut off. It is then switched into conduction during one half of the input cycle. Two devices are necessary so that both halves of the cycle can be dealt with, each device amplifying alternate half-cycles. This class of operation is usually reserved for large power audio and radio-frequency amplification where it has the advantage of a much greater efficiency than Class A working.

Class C. The transistor or valve is biased beyond the current cut-off point so that the input signal has to exceed the relatively high value of applied bias before the device switches on. This means that the output signal consists of a series of short current pulses. This class of operation is reserved for tuned oscillators and some radio-frequency power amplifiers. Its efficiency is much greater than either Class A or Class B.

Other classifications are possible, such as Class AB which we will discuss later, but at this stage we shall be interested only in Class A operation.

The common-emitter amplifier

The common-emitter configuration is the most widely used circuit arrangement and the basic single-stage amplifier for this mode of working is shown in *Figure 2.6.* We shall consider an *n-p-n* transistor throughout the following sections, but the general discussion and theory is equally applicable to *p-n-p* devices.

The input signal to the amplifier is developed between the base and

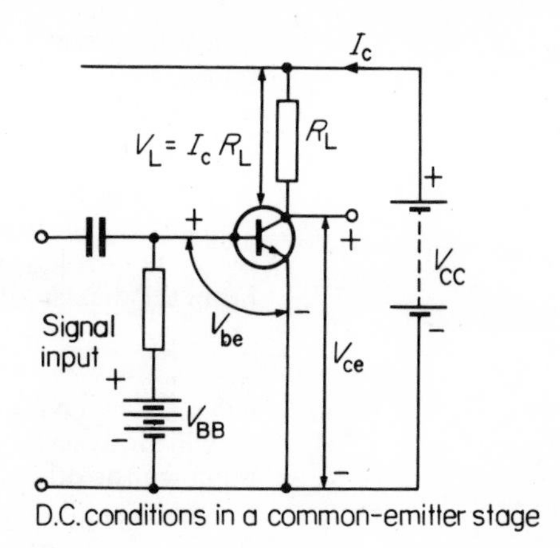

Figure 2.6

emitter terminals in series with the base bias voltage V_{BB} (here shown as a battery supply for convenience of description). This bias voltage has its polarity arranged so that the base-emitter junction diode is conducting, that is, the diode is forward biased. The output signal is developed in turn across the collector load resistor R_L which (as far as the a.c. signal is concerned) is effectively wired between collector and common earth line. The magnitude of the direct current established by the supply V_{BB} in the base-emitter circuit determines the class of operation of the transistor; for the amplifier we are discussing this will be Class A, collector current flowing for the whole of the input cycle.

The current or voltage gain of such a resistance loaded amplifier can be determined graphically with the aid of a load line drawn across the output characteristics of the transistor. These characteristics relate the

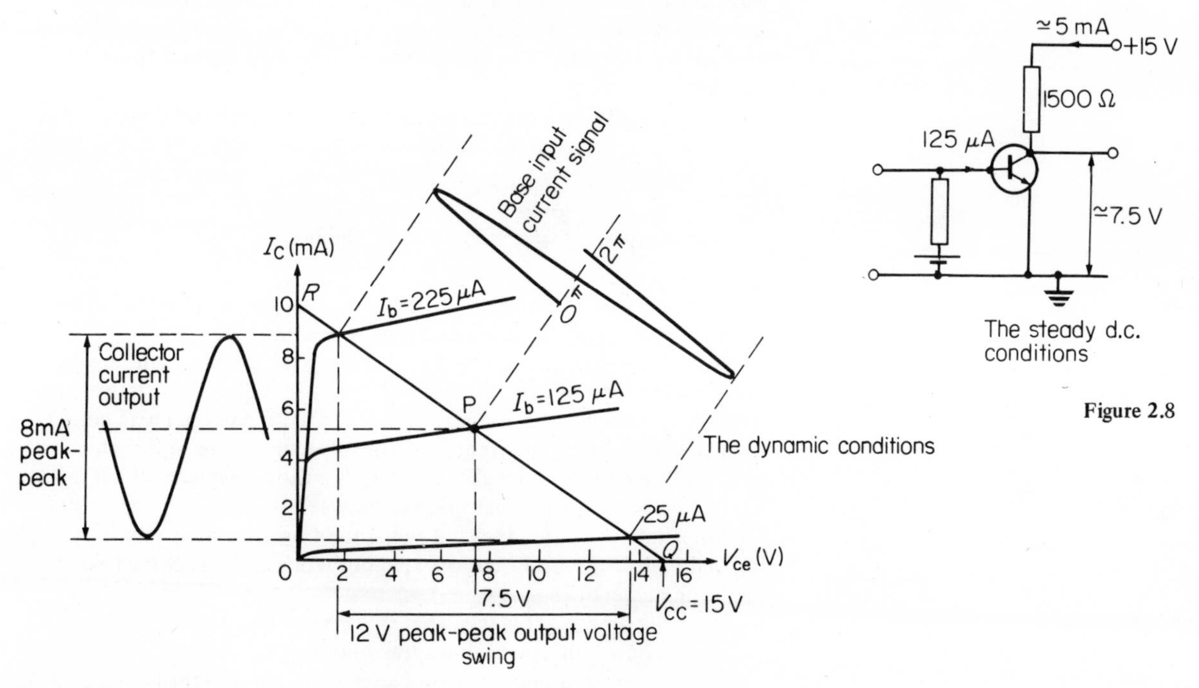

Figure 2.8

Figure 2.7

collector current I_C to the collector voltage, V_{ce}, for various fixed values of base current I_b, and a typical family is shown in *Figures 2.7 and 2.8*. To determine the position of the load line we make use of the d.c. conditions illustrated in *Figure 2.6*.

The collector current flowing in the load resistor R_L develops a voltage equal to $I_C R_L$ across it, and the direct voltage applied across the transistor, V_{ce}, is the supply voltage V_{cc} minus this voltage drop. From the diagram the following relationship clearly follows:

$$V_{cc} = V_{ce} + I_C R_L \qquad (2.1)$$

Since V_{cc} and R_L are fixed for a given circuit, the variables V_{ce} and I_C will always bear a definite relationship to each other so that if one is known the other can be calculated. The load line which relates these variables for all their possible values can now be drawn across the characteristic curves in the manner illustrated. The supply voltage V_{cc} determines the value of the collector voltage V_{ce} when the collector current I_C is zero. This condition establishes the position of point Q, for at $I_C = 0$ the transistor is cut off and clearly $V_{ce} = V_{cc}$.

In the diagram we have assumed that V_{cc} is 15 V, so the load line intersects the V_{ce} axis at this point. Point R at the other end of the line is established as that value of collector current which flows when V_{ce} is assumed zero, that is, when the whole of the supply is assumed to be dropped across R_L and the transistor is saturated (or bottomed). From equation (2.1) above, this value of I_C is seen to be V_{cc}/R_L, and since the value of R_L is taken to be 1500 Ω the load line intersects the collector current axis at $I_C = 15/1500\text{A} = 10$ mA. Since $V_{ce} = V_{cc} - I_C R_L$ the load line equation is that of a straight line, so by joining Q and R with a straight line, the required load line for the case where $R_L = 1500$ Ω is established.

Load lines for other values of R_L can be drawn in exactly the same way, or of course for other values of V_{cc}. Notice that the load line traverses the characteristics from the region of saturation to the cut-off limit of the transistor; the line cannot be extended beyond these extremes.

The gradient of a load line is important and is measured as the ratio of the current axis magnitude to the voltage axis magnitude at the limits of the line, that is

$$\text{gradient} = \frac{-V_{cc}}{R_L} \div V_{cc} = -\frac{1}{R_L} \text{ siemen}$$

the negative sign indicating that the gradient is greater than 90°. The gradient of the line decreases as R_L increases and vice versa.

The effect of an alternating signal at the base terminal can now be analysed in terms of the load line and the output characteristics of the transistor. For with the operating point at P (the no-signal bias point) the excursions of base current I_b permissible about this point which will result in *proportional* changes in I_C are those between which the load line extends over equal spacings between the static characteristics on either side of P, and does not swing the transistor at any time beyond collector current cut off at Q nor work into the curved portions of the characteristics at R.

From the diagram it is seen that this maximum excursion of base current is about 200 μA peak-to-peak about an operating level of 125 μA. The corresponding excursion in collector current is from 1 mA to 9 mA, a peak-to-peak swing of 8 mA about a mean current of 5 mA. The swing in collector voltage is then of the order of 12 V peak-to-peak.

The current gain of the amplifier can now be calculated, for with a 200 μA peak-to-peak input signal, the output current is 8 mA peak-to-peak. Hence

$$A_i = \frac{\delta i_O}{\delta i_i} = \frac{8 \times 10^{-3}}{200 \times 10^{-6}} = 40$$

The choice of operating point P for Class A operation is usually somewhere about the centre of the load line but must be selected in practice with regard to the above mentioned statements about the permissible input base current excursion. The effect of a badly selected operating point will become apparent when temperature problems are discussed in the section 3. This is a reason for expressing the current gain in terms of the ratio of peak-to-peak values; the two half-cycles of a distorted waveform will have different peak values, so by using peak-to-peak figures a more accurate result is obtained.

Example (5) A transistor connected in common-emitter mode has the data given in the following table:

	I_C (mA)			
V_{ce} (V)	I_b = 20 μA	40 μA	60 μA	80 μA
3	0.91	1.6	2.3	3.0
5	0.93	1.7	2.5	3.25
7	0.97	1.85	2.7	3.55
9	0.99	2.05	3.0	4.05

Plot the output characteristics for the given base currents and use these characteristics to determine (a) the current gain when V_{cc} is 6 V, (b) the output resistance of the transistor for I_b = 60 μA.

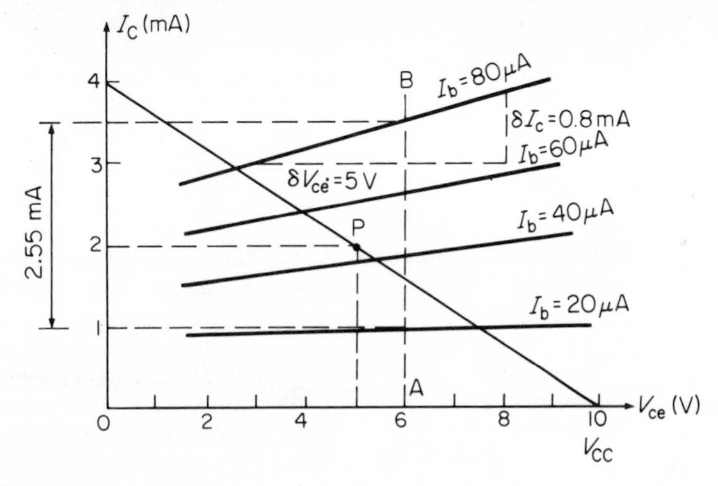

Figure 2.9

The transistor is to be used with a load resistor of 2500 Ω and a collector supply of 10 V. Draw the load line for this condition and use it to find the base current for a collector voltage of 5 V.

The required characteristic curves are plotted in *Figure 2.9*. To plot the load line, two points are needed to define the extremities of the line. One point is found by setting $I_C = 0$, for then $V_{ce} = V_{cc} = 10$ V and this is marked Q in the diagram. The other point is $V_{ce} = 0$ and then $I_C = V_{cc}/R_L$; this gives point R at $I_C = 4$ mA.

For parts (a) and (b) of the problem the load line is not required.

(a) When $V_{ce} = 6$ V, a change of (80–20) μA = 60 μA in base current gives a change of (3.5–0.95) mA = 2.55 mA in collector current. This is found by erecting a vertical from the $V_{ce} = 6$ V point on the horizontal axis to cut the characteristics at A and B; by projecting across to the current axis the corresponding change in I_C is noted. Hence

$$A_i = \frac{2.55 \times 10^{-3}}{60 \times 10^{-6}} = 42.5$$

This figure is the *static* gain of the transistor; the *dynamic* or true signal gain would differ somewhat from this value.

(b) The output resistance is expressed by the ratio of the change in output voltage to the change in output current, that is $R_O = \delta V_{ce}/\delta I_C$ and so is given by the gradient of the 60 μA characteristic concerned. From the triangle drawn on the curve

$$R_o = \frac{5}{0.8 \times 10^{-3}} = 6.25 \text{ k}\Omega$$

The last part of the question involves the load line; when $V_{ce} = 5$ V the operating point will be where the vertical projection from this value cuts the load line:- point P. The base current now has to be estimated from the position that P occupies between $I_b = 40$ μA and 60 μA. From inspection, I_b is approximately 45 μA. This would be about the right bias current for this load line.

(6) We obtained the static current gain figure of 42.5 in the above problem. Assuming that we are working at point P, and the input signal swings about this value (45 μA) with a peak amplitude of 20 μA, estimate (a) the peak-to-peak collector current swing, (b) the dynamic current gain.

Setting the operating point

Return now to the circuit diagram of *Figure 2.6* and the characteristic curves of *Figure 2.7 and 2.8*. The base-emitter voltage necessary to set up the required base bias current (in this example 125 μA) is obtained in practice not from a separate battery supply but from the collector supply rail. The battery V_{BB} which supplies forward bias to the base has the same polarity, with respect to the emitter, as the battery V_{cc} which supplies reverse bias to the collector. Therefore, battery V_{cc} can supply the proper bias to both junctions, the base being returned to

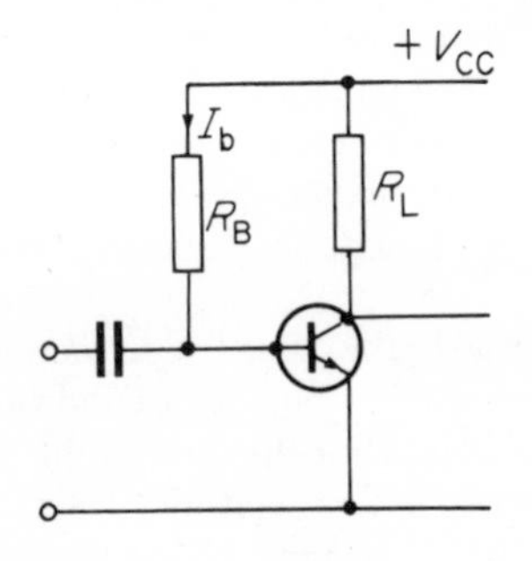

Figure 2.10

the V_{cc} rail by way of a bias resistor R_b as shown in *Figure 2.10*. The value of this bias resistor is easily calculated by Ohm's law:

$$R_b = \frac{V_{cc} - V_{be}}{I_b}$$

If V_{cc} is large in comparison with V_{be} this will simplify to $R_b = V_{cc}/I_b$. For the example of *Figure 2.8*, where V_{cc} = 15 V and the base current is 125 μA, R_b is then approximately $15/125 \times 10^{-6}$ = 120 kΩ. This value actually includes the internal base-emitter resistance, but as this is of the order of only a few hundred ohms at most it may be neglected. It is important to notice that the bias is *not* developed across R_b but across the base-emitter junction as a result of the no-signal flow of current through that junction. This action makes the base positive with respect to the emitter and the diode is forward biased. The direction of the voltage drop across R_b itself is *opposite* to this forward bias polarity.

Bias of this simple form is *fixed* bias because the bias current I_b is determined almost entirely by the values of R_b and V_{cc}. It might seem because R_b and V_{cc} would be constant in a particular circuit that the operating point determined by this form of bias would also be constant. However, the characteristics of transistors are functions of temperature, and the thermally generated leakage current I_{CBO} increases as temperature increases.

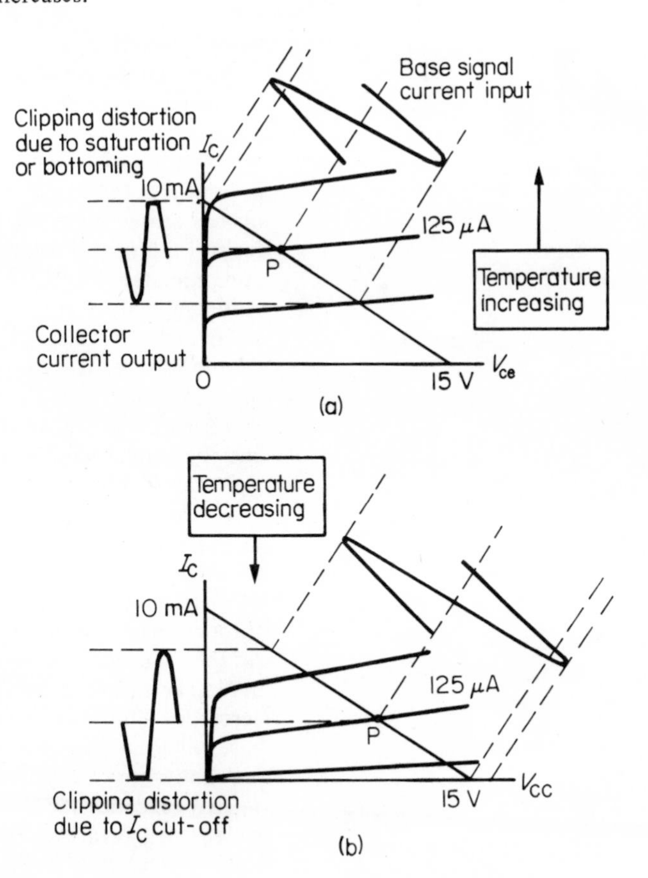

Figure 2.11

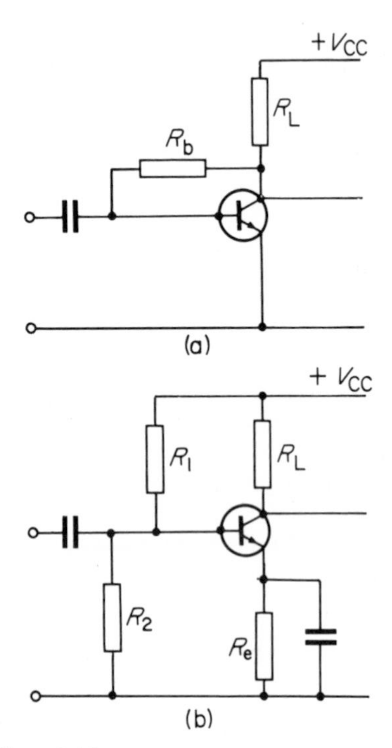

Figure 2.12

When the temperature changes, the characteristics move vertically across the axes, upwards for an increase in temperature and downwards for a decrease in temperature, and the spacing between the curves also changes. This action, superimposed on to the existing axes and load line background will effectively shift the operating point P one way or the other along the load line.

Figures 2.11(a) and *(b)* show respectively the effect of an increase and a decrease in ambient temperature upon the operating conditions of the circuit previously discussed. The reason for the increase in collector current which follows from a rise in temperature (the effect of most importance) even though I_b is held constant, is that I_{CBO} is forced to flow across the emitter junction and, as we noted in the Electronics course 2, is therefore amplified by the factor (α_E +1), where α_E is the static current gain in common-emitter mode. As soon as the operating point P moves appreciably towards the upper end of the load line, the amplifier becomes useless. Further, as I_C increases still further, the transistor heats up internally so aggravating the situation and leading finally to thermal runaway and destruction of the device.

The single bias resistor circuit is satisfactory only where the ambient temperature is likely to be constant and the transistor is operated at such low power levels that any internal heating is negligible. The circuit also suffers from the disadvantage that the transistor cannot be changed for another unless the replacement is closely matched to the original, for $I_C = I_{CBO} + \alpha_E I_b$ and the variation in α_E is considerable over a batch of transistors.

Some improvement can be obtained by connecting the bias resistor between collector and base as shown in *Figure 2.12(a)*. With this arrangement R_b provides self-bias for the emitter junction and is in series with the bias-emitter circuit. Referring to the example of *Figures 2.7 and 2.8*, the quiescent collector voltage was 7.5 V and the required bias current was 125 μA. For this current to flow in R_b when this resistor is returned to the collector requires that R_b must now be approximately $7.5/125 \times 10^{-6}\ \Omega = 60\ k\Omega$ or half its previous value. Any temperature produced increase in collector current results in a fall in collector voltage and hence to a decrease in the base bias current. This in turn reduces the collector current, so compensating for the original change.

Figure 2.12(b) shows the most commonly used bias arrangement, and this provides stability with respect both to temperature variation and transistor interchangeability.

A resistor R_e is now included in the emitter lead and the base is fed from the centre connection of a potential divider circuit made up of resistors R_1 and R_2. If this potential divider is designed to have good regulation, that is, if R_2 is small in value and the total current flowing through the divider is very large in comparison with the base current I_b, then the base voltage V_b will be held constant in spite of variations in I_b. The emitter is held at a positive potential V_e by the voltage drop across R_e. The base-emitter voltage V_{be} is the difference between the base voltage V_b and V_e and the resistor values are chosen so that the junction is forward biased by a fraction of a volt.

Any tendency for the collector (and hence the emitter) current to increase will create an increased voltage drop across R_e and this will lead to a reduction in V_{be}. The base current will be reduced accordingly causing a decrease in the collector current that compensates for the original increase. Make a note here that in the emitter-biased circuit system, a constant level of *emitter current* is being set, rather than constant base current.

In theory R_e should be large enough and R_1 and R_2 small enough to give a base potential of effectively zero d.c. resistance. However, R_e cannot be too large or a prohibitive proportion of the available supply voltage V_{cc} will be lost across it; and the potential divider resistance must not be so low as to load the input signal source severely (R_1 and R_2 are effectively in parallel with the signal source from an a.c. point of view), or to draw an excessive current from the supply.

It is usual to make R_e of such a value that a voltage drop of about 1 V occurs across it and to proportion the divider so that R_2 is about five to ten times the value of R_e. The total bleed current through the divider is also set to be about ten times the required bias current.

(7) In the bias circuit arrangement illustrated in *Figure 2.12(a)*, is a constant I_b being set?

(8) In *Figure 2.12(b)* a capacitor is shown connected in parallel with R_e. Can you explain at this stage why this is necessary?

Example (9) The common-emitter amplifier of *Figure 2.13* has a V_{cc} supply of 15 V and requires a base bias current of 100 μA. If the base-emitter voltage is then 0.3 V and the quiescent emitter current is 2 mA, assess suitable values for resistors R_1, R_2, R_e and R. Which preferred values would you choose in practice?

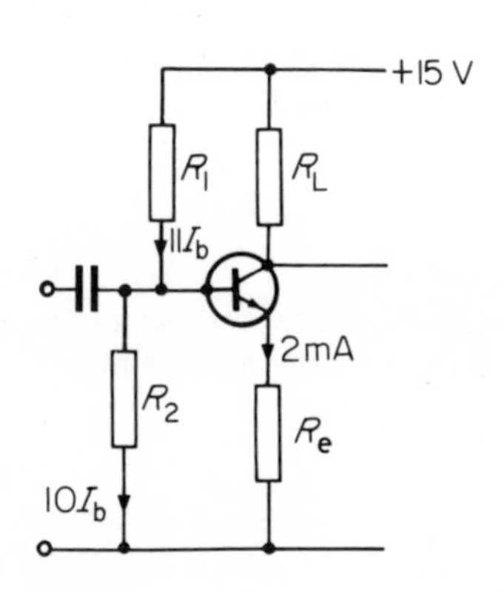

Figure 2.13

We have to *assess* suitable values, not attempt to find unique values for the circuit resistances. Assume, therefore, that a voltage drop of 1 V is reasonable across R_e. Then

$$R_e = V_e/I_e = 1/(2 \times 10^{-3}) = 500\ \Omega$$

A 470 Ω preferred value could be used here.

The current through the divider has to be large compared with I_b which is 100 μA; let the current through R_2 be 10 I_b or 1 mA, then the current through R_1 will, by Kirchhoff, be 11 I_b or 1.1 mA. From the circuit, the voltage across $R_2 = V_{be} + I_eR_e$

$$= 0.3 + 1.0 = 1.3\text{ V}$$

But this is also equal to $1 \times 10^{-3} \times R_2$, so that

$$R_2 = \frac{1.3}{10^{-3}} = 1.3\text{ k}\Omega$$

A 1.5 kΩ preferred value could be used here.

The current in R_1 is 1.1 mA, and the voltage across R_1 is $V_{cc}-1.3 = 15-1.3 = 13.7$ V. Hence

$$R_1 = \frac{13.7}{1.1} \times 10 = 12.45\text{ k}\Omega$$

A 12 kΩ preferred value could be used here.

Finally, since there is a 1 V drop across R_e, 14 V is available for division between R_L and V_{ce}. This may be divided equally, so giving a mean collector voltage of 7 V. Hence, since $I_C \simeq I_e$

$$R = \frac{7}{2} \times 10 = 3.5 \text{ k}\Omega$$

A 3.3 kΩ preferred value could be used here.

The A.C. load line

It is not often that the load line on which a transistor operates is the one drawn for d.c. conditions as discussed above. The reason for this is that the load into which the transistor works is different for a.c. (signal) and d.c. (static) conditions. Two load lines are then required for a proper examination of the circuit: the d.c. load line which establishes the proper operating point, and an a.c. load line from which the actual current or voltage gain of the amplifier may be estimated.

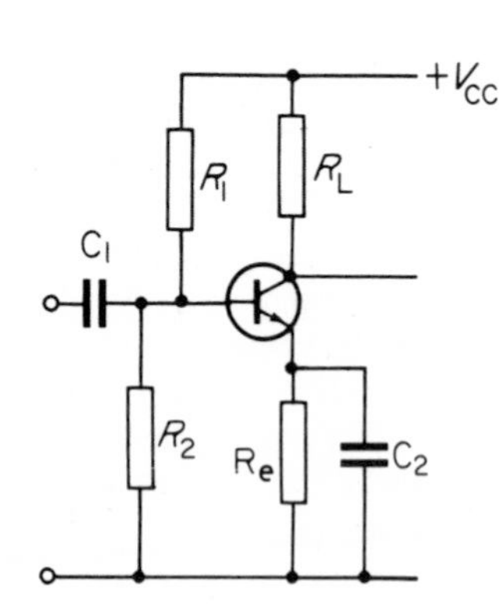

Figure 2.14

Figure 2.14 shows a single stage amplifier with emitter resistance R_e and a potential divider bias circuit. R_e is by-passed by a large value capacitor C_2 which effectively short-circuits the resistance at the signal frequency but not at d.c. (or zero frequency). At signal frequencies, therefore, the effective load on the transistor is only R_L and

$$V_{cc} = V_{ce} + I_c R_L$$

which is the same as the equation obtained earlier when the emitter was assumed to be connected directly to the earth line. This equation now represents the a.c. load line and its gradient is $-1/R_L$ siemen.

At zero frequency, C_2 is ineffective in shunting R_e and the d.c. load on the transistor is $(R_L + R_e)$ so that the equation of the d.c. load line becomes

$$V_{cc} = V_{ce} + I_c(R_L + R_e) \quad (2.2)$$

This line has a gradient of $-1/(R_L + R_e)$ siemen and this will clearly be *less* than the gradient of the a.c. line. Both lines are shown on the characteristics of *Figure 2.15.*

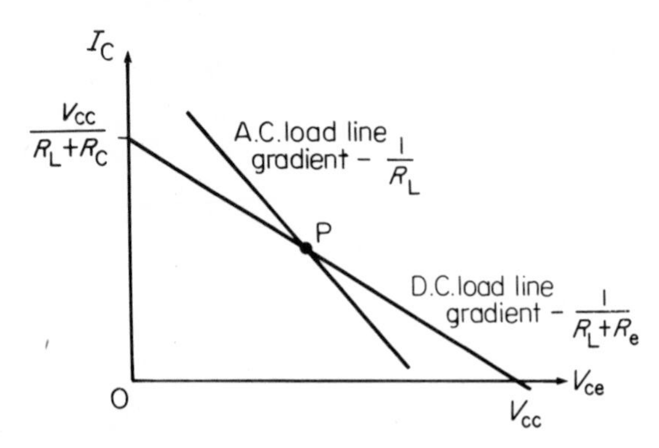

Figure 2.15

To establish the proper operating conditions now, the d.c. load line is drawn first and a suitable operating point selected at P. The a.c. load line is then drawn, *passing through* P and having a gradient equal to $-1/R_L$. It is particularly important to take note of the italicised words here: the a.c. load line *must* pass through P and is *not* established (as is the d.c. load line) by intersections on the respective axes.

In many cases, especially where R_e is small in comparison with R_L, it is sufficiently accurate to use the single a.c. load line and select the operating point directly on that.

Example (10) In *Figure 2.14* $R_L = 4.7\ k\Omega, R_e = 1\ k\Omega$ and $V_{cc} = 10$ V. Sketch the d.c. and a.c. load lines on appropriate axes.

The gradient of the d.c. load line is $-1/(R_L + R_e) = 1/5700$ S. This line will cut the horizontal axis at the point $V_{ce} = V_{cc} = 10$ V, and the vertical axis where $I_C = V_{cc}/(R_L + R_e) = 10/5700$ A $= 1.75$ mA as shown in *Figure 2.16*. Taking an operating point P at the centre of this line, the a.c. load line must be drawn passing through P with a gradient equal to $-1/R_L = 1/4700$ S. A convenient way of doing this is to draw a 'dummy' line of the required gradient between two convenient points on the axes: assume that we are dealing with a d.c. load line, then using some handy value of supply voltage, say 4 V as shown on the diagram, calculate the corresponding value of I_C (for $V_{ce} = 0$ from equation (2.2) above) i.e. $4/4700$ A $= 0.85$ mA. This establishes the two extremes of the 'dummy' line and hence it will have the required gradient of the wanted a.c. load line. The true a.c. load line can now be drawn parallel to the dummy and passing through P.

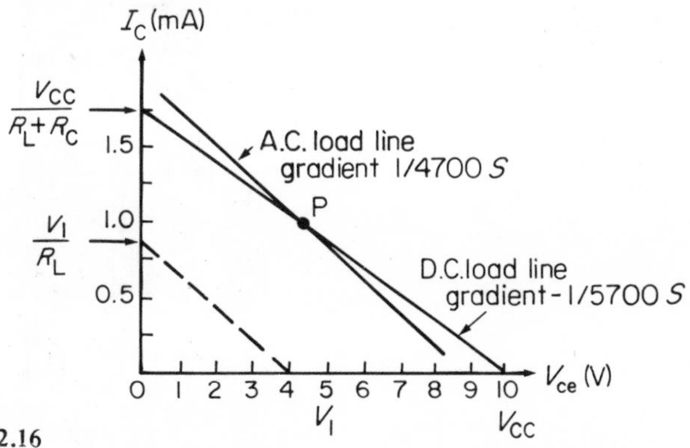

Figure 2.16

COUPLING METHODS

You may have noticed throughout the previous discussion that a capacitor, such as C_1 in *Figure 2.14*, was included in the base input lead of all the amplifier circuits. This capacitor did not enter into the discussions on the various d.c. circuit conditions because, being an equivalent of an open-circuit to direct currents, its presence in no way affected, for example, the location of the load line or the operating point. To an a.c. signal applied at the input terminals, however, the capacitor would have a small reactance so that the signal voltage would be passed on and applied to the base terminal of the transistor without in any way disturbing the carefully set-up d.c. conditions.

When a signal input is applied to a transistor, or the output of one transistor is applied as input to the next, some form of coupling is required which permits the signal to pass easily between transistors but which at the same time prevents the applied d.c. voltages from being disturbed. *Direct* coupling is possible and is frequently used, but this is a particular case which will be separately discussed later on.

There are several methods of connecting one amplifier stage to another so that the signal output of the first is applied as input to the second. The circuits making up such connections are known as *interstage couplings.* As we have just mentioned, one obvious form of coupling is by way of a capacitor, this component allowing the signal frequencies to be transferred from one stage to the next relatively unhindered but acting as an open-circuit to the d.c. potentials on the transistor terminals. The coupling does not have to take this form, but may be made by way of a transformer alone or by a transformer in association with a capacitor. Each method has its own particular advantages and disadvantages and the actual choice of coupling depends upon the performance specifications (as well as the cost) expected from the completed amplifier.

Equivalent circuits

Multistage amplifiers are best studied in terms of *a.c.equivalent circuits.* So far we have been concerned only with the d.c. operating conditions of single-stage amplifiers, but if we assume that these d.c. conditions are properly set, then we need consider only the a.c. signal voltages and currents for a dynamic analysis of the circuit.

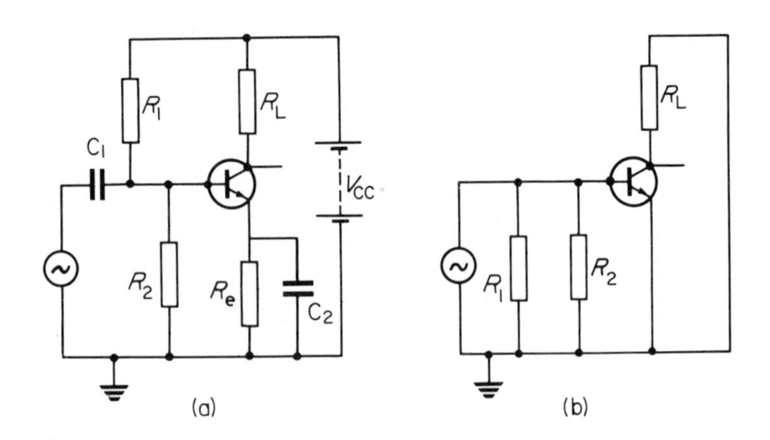

Figure 2.17

Look at *Figure 2.17(a)* which shows a common-emitter amplifier with the d.c. supplies and biasing appropriately connected. If we wish to draw an a.c. equivalent circuit of this amplifier, we obtain the diagram shown at (*b*). Here the d.c. supply battery V_{cc} has disappeared, along with resistor R_e, and R_1 has turned up in parallel with R_2 across the base input terminals. In addition, the collector load R is now effectively wired between the collector and the earth line. This metamorphosis has come about because *to the signal* the battery has zero internal resistance and the emitter capacitor C_2 has zero reactance.

Also, with V_{cc} short-circuited to a.c. R_1 'bends' round to be in parallel with R_2, and R is carried round to the earth line. So diagram (b) tells us what the circuit looks like at the signal frequencies. The V_{cc} supply simply sets the proper operating conditions and enables the current carriers to get to work inside the transistor; it is of no interest in an analysis of the circuit under signal conditions.

In your later advanced work you will find the use of equivalent circuits of the greatest value, and certainly necessary in amplifier analysis.

Resistance-capacitance coupling

This is a very common method of coupling and requires only inexpensive components. The basic circuit arrangement is shown in *Figure 2.18*.

Bias for both stages is provided by potential dividers R_1R_2 and R_5R_6 and emitter resistors R_3 and R_7 are bypassed with large value capacitors C_3 and C_4 respectively. The collector load resistors are R_4 and R_8.

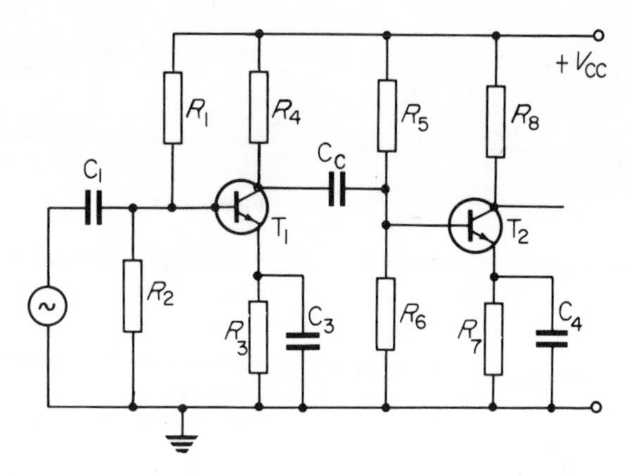

Figure 2.18

We are interested in the coupling between transistors T_1 and T_2, so the components of immediate concern are R_4, C_C, R_5 and R_6 and this part of the circuit is transferred to *Figure 2.19(a)* where the load resistor has been given its usual designation of R_L. The output voltage of T_1 developed across R_L is coupled to the base of T_2 via capacitor C_C, the signal input to T_2 being developed across R_6. Now what does the circuit look like to the signal voltage? Well, the V_{cc} line is 'dead' to this voltage and can be effectively shorted down to the earth line. So the a.c. equivalent circuit becomes as shown in *Figure 2.19(b)*.

As a transistor behaves as a constant current generator we can represent T_1 as such a generator sending current into the load resistor R_L and developing a voltage V_i across it. The coupling capacitor which we have for the present left as C_C couples the signal voltage developed across R_L into the base of T_2 which is now effectively shunted by the parallel arrangement of R_5 and R_6. The output voltage V_o from the coupling circuit is the input signal to T_2.

A further simplification is now possible, for by combining the parallel circuit of R_5, R_6 and the input resistance of T_2 into a single resistance R_i, the equivalent circuit reduces to that shown in *Figure 2.19(c)*. There is one capacitor here we have not so far encountered, C_S shown in parallel with R_L. This component represents the circuit stray capacitance which is the lumped sum of the output capacitance of T_1, the input capacitance of T_2 and the stray wiring capacitances across R_L.

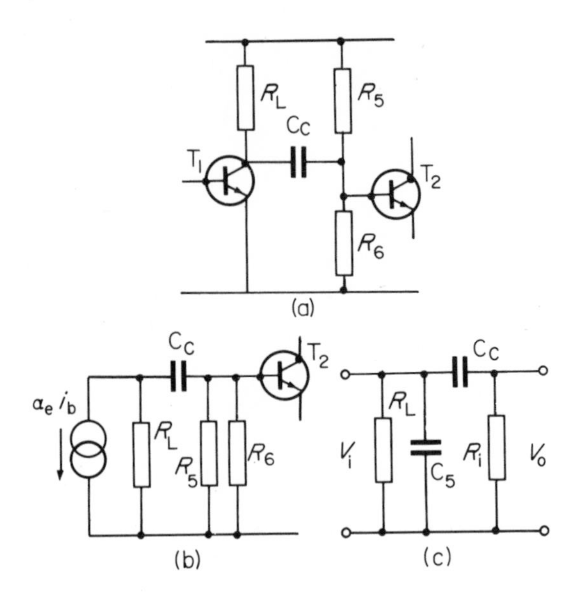

Figure 2.19

The way the coupling behaves in transferring the signal from T_1 to T_2 depends upon the presence of capacitor C_C and the stray capacitance C_S. For convenience of analysis we divide the frequency range of the signals that the amplifier is designed to amplify into three sections: low-frequency, mid-frequency and high-frequency sections. You should now refer back to *Figure 2.3(a)* to refresh your memory on this. The term 'mid-frequency range' is used to define that part of the frequency range in which the effect of all *reactances* is negligible and the gain is consequently independent of frequency.

The low- and high-frequency ranges are those parts of the frequency range in which reactances cause the gain to become dependent upon frequency. So, in the simple equivalent circuit of *Figure 2.19(c)*, coupling capacitor C_C, being of relatively large capacitance, is assumed to have *zero reactance* in the mid-frequency range, while capacitor C_S, being of relatively small capacitance, is assumed to have a reactance so large that its parallel effect on the circuit is negligible.

In the mid-frequency range therefore, both capacitors can be ignored so that the remaining circuit, being made up only of resistive elements, has transmission properties which are independent of frequency. The response curve consequently stays flat over the mid-frequency range.

One further point is that there will be no phase shift through the circuit, hence V_o will be in phase with V_i.

Low- and high-frequency effects

Consider now the behaviour of the coupling circuit at low frequencies. The reactance of the shunting stray capacitance C_S will be very high relative to the parallel R_L and may be neglected, just as it is at mid-frequencies. The coupling capacitor C_C, however, will now exhibit a reactance which is comparable with or larger than the resistances which 'face' it*, hence the equivalent circuit reduces to that shown in *Figure 2.20*.

*The resistance facing a capacitor is the resistance encountered in moving round the circuit from one capacitor plate to the other.

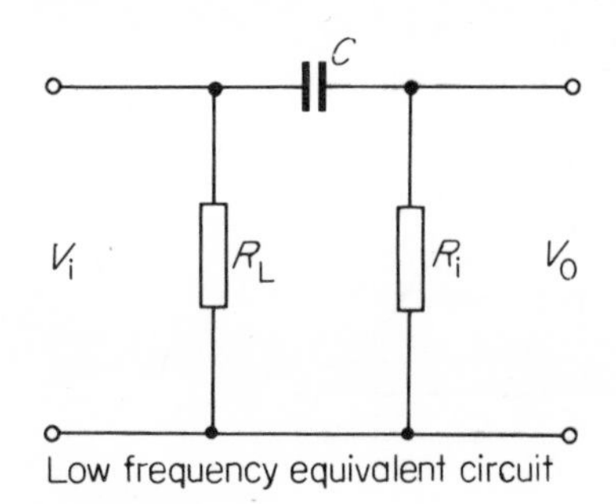

Figure 2.20

At some frequency f_1 when the reactance of C_C becomes equal to the total resistance facing it, the output voltage V_o will be equal not to V_i but only to 0.707 V_i. This reduction, you will recall, represents the half-power point level on the response characteristic; and the output voltage will progressively reduce as the frequency decreases still further. Hence at low-frequencies the gain falls from its mid-frequency level as the earlier response curve indicated.

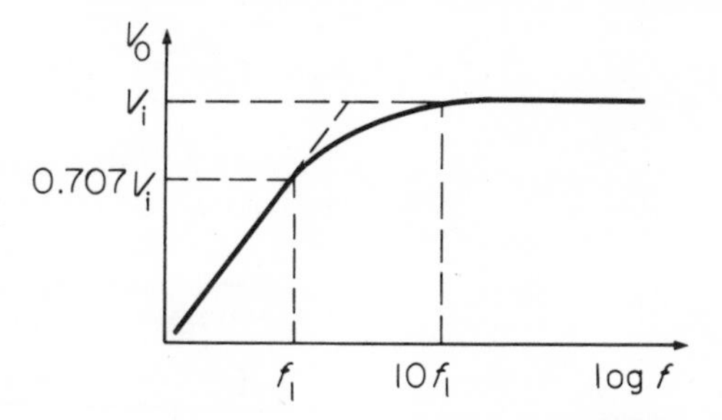

Figure 2.21

A plot of gain versus frequency for this part of the spectrum is given in *Figure 2.21*, and it is seen that the gain is closely proportional to frequency. As the frequency increases, the gain approaches the mid-band value and at the half-power point where $f = f_1$

$$R_L + R_i = \frac{1}{2\pi f_1 C_C}$$

It is common practice to assume that the reactance of C_C can be neglected if it is less than one-tenth as large as the facing resistance, so that if $X_{cc} = 0.1(R_L + R_i)$ the frequency will be $10f_1$. At this frequency the gain is only about 0.5% below its mid-band value and so the frequency at which the low-frequency region is considered to be at its upper limit is $10f_1$.

(11) In the circuit of *Figure 2.18*, $R_L = 10\ \text{k}\Omega$, $R_5 = 15\ \text{k}\Omega$, $R_6 = 3\ \text{k}\Omega$ and $C_C = 1\ \mu\text{F}$. If the input resistance of transistor T_2 is 2 kΩ, what will be the low-frequency half-power point on the response curve of this amplifier?

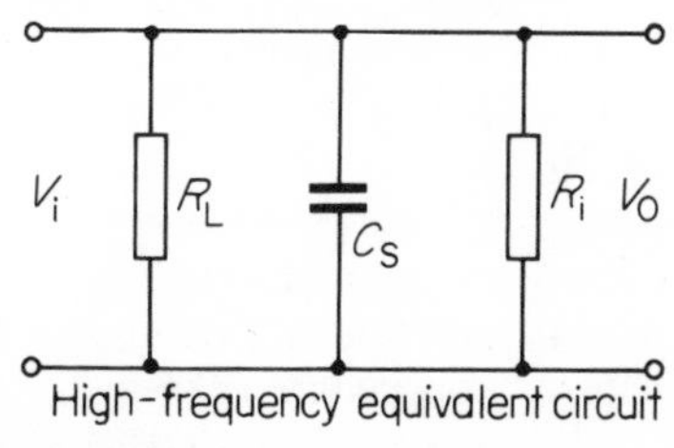

Figure 2.22

Now consider the behaviour of the coupling circuit at high frequencies. As for the mid-frequency range the coupling capacitor C_C will have zero reactance and may be ignored. C_S however will have a reactance which also decreases as the frequency increases and because of its position it will increasingly shunt the collector load R_L, so reducing the available output voltage developed across R_i and feeding into T_2. The equivalent circuit now reduces to that shown in *Figure 2.22*. At some frequency f_2 when the reactance of C_S becomes equal to the total resistance facing it (R_L and R_i in parallel), the input to T_2 will be reduced to 0.707 of its maximum value at mid-frequencies, and this will progressively reduce as the frequency increases further.

A plot of gain versus frequency for this part of the spectrum is given

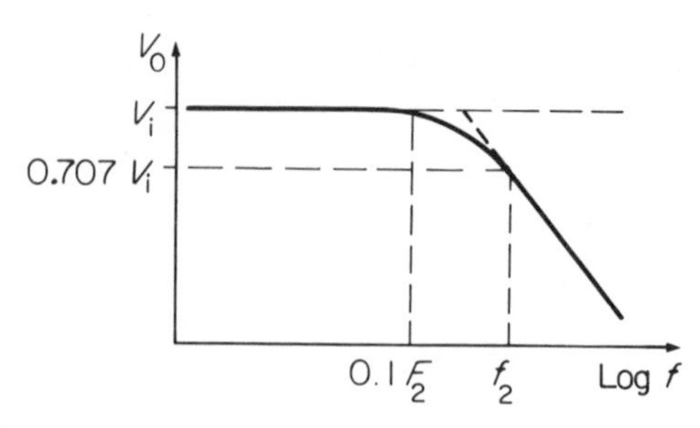

Figure 2.23

in *Figure 2.23* and again it is seen that the gain is closely proportional to frequency. As the frequency falls, the gain approaches the mid-band value and at the half-power point where $f = f_2$

$$\frac{R_L R_i}{R_L + R_i} = \frac{1}{2\pi f_2 C_S}$$

It is usually accepted that the gain becomes equal to the mid-band value at a frequency given by $0.1f_2$. This is indicated on the diagram.

Circuit components

From what we have just discussed it follows that the bandwidth of an R-C coupled amplifier will be large if (i) the circuit strays and inter-electrode capacitances are kept small, and (ii) if the coupling capacitor value is large. The former condition can be achieved by careful attention to the circuit layout. Because the input resistance of a common-emitter stage is low, the effective value of R_i in the equivalent circuit representation is also low, hence a very large value of coupling capacitor is necessary if the gain is to be maintained down to very low frequencies. Electrolytic capacitors are normally used, and values of some hundreds of microfarads are not unusual.

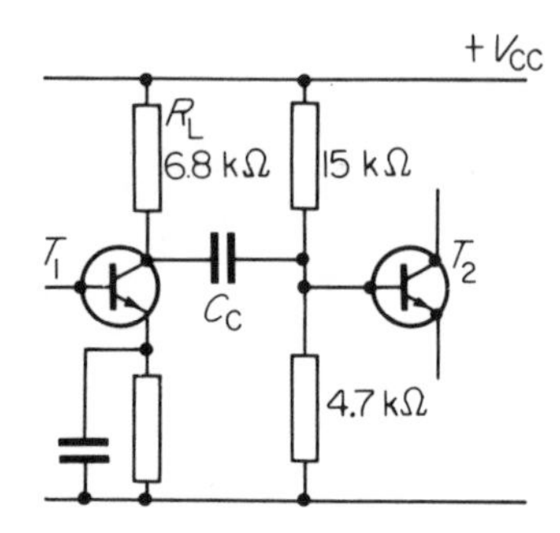

Figure 2.24

(12) In the circuit of *Figure 2.24* the input resistance of transistor T_2 is 1 kΩ. What value of coupling capacitor will be required if the 3 dB point on the response curve is not to be greater than 25 Hz?

Example (13) In order to limit the high frequency response of the amplifier of the previous example, a capacitor is connected in parallel with the collector load resistor of T_1. Why will this limit the response? If the stray capacitance add up to a total effective value of 50 pF already, what additional capacity is required in this position to reduce the upper half-power point to 10 kHz?

The added capacitor will effectively increase the stray capacitance C_S and so reduce the upper frequency response for the reason already explained in the text.

At the upper 3 dB or half-power point, we have from the equation above

$$\frac{R_L R_i}{R_L + R_i} = \frac{1}{2\pi f_2 C_S}$$

This time, C_S will be made up of the existing stray capacitances plus the added capacitor. Working out the effective shunt resistance on the left hand side we have for R_i for the 15 kΩ, the 4.7 kΩ and the 1 kΩ input resistance of T_2 all in parallel, hence

$$\frac{1}{R_i} = \frac{1}{15} + \frac{1}{4.7} + \frac{1}{1} \text{ mS}$$

and from this

$$R_i = 0.78 \text{ k}\Omega$$

Hence

$$\frac{R_L R_i}{R_L + R_i} = \frac{6.8 \times 0.78}{7.58} = 0.7 \text{ k}\Omega \ (700\ \Omega)$$

Then

$$700 = \frac{1}{2\pi \times 10 \times 10^3 \times C_S}$$

and

$$C_S = \frac{1}{2\pi \times 10 \times 10^3 \times 700} \text{ fF}$$

$$= \frac{10^{12}}{2\pi \times 10 \times 10^3 \times 700} \text{ pF}$$

$$= 22736 \text{ pF } (= 0.0227\ \mu\text{F}).$$

The required added capacitance will be therefore 22736 – 50 = 22686 pF. This example shows that the shunting effect of capacitance across the collector load is very small upon the high-frequency limit of an amplifier and for normal stray capacitances its effect is negligible at audio frequencies.

Phase conditions in the R-C coupling

At the mid-frequency range, we have seen that the phase shift through the coupling is zero. Hence the *total* phase shift from the input of transistor T_1 to the input of transistor T_2 is 180°, this shift being the sum of the shift in the transistor itself (180°) plus the shift in the coupling (0°). We adopt the convention that in the transistor, the output voltage leads the input voltage.

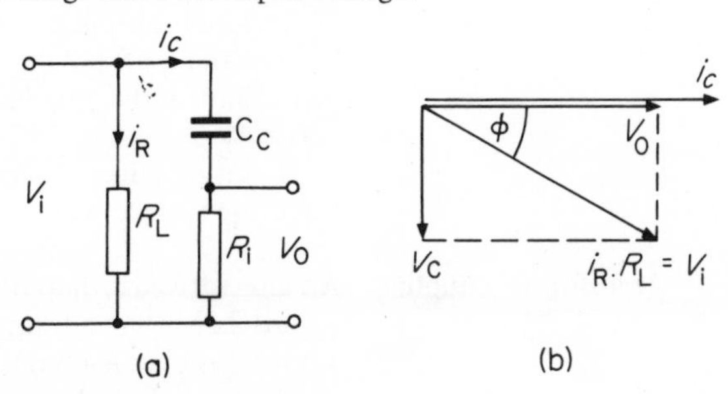

Figure 2.25

At low frequencies, when the circuit is equivalent is as shown in *Figure 2.25(a)* the total phase shift is no longer 180°. Ignoring the 180° transistor shift for the moment and considering only the coupling network, we can sketch the phasor diagram of *Figure 2.25(b)*. The voltage V_C across C_C lags the current i_C by 90°, and v_O is in phase with i_C (since this current also flows in R_i). The resultant of these two voltages is equal to the applied voltage developed across R_L, which is the input voltage v_i. Hence the output voltage leads on the input voltage by some angle ϕ which can lie between 0° and 90°, the greatest possible value of 90° being approached as the frequency becomes very small.

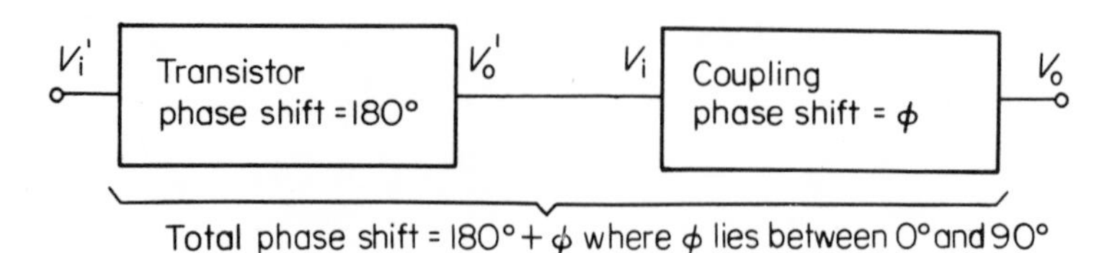

Figure 2.26

The overall phase shift can now be evaluated from a consideration of *Figure 2.26*. In the transistor, v_o leads v_i' by 180°. In the coupling circuit, v_o leads v_i by an angle lying between 0° and 90°. Hence the total shift is 180° + (angle between 0° and 90°) which is some angle between 180° and 270°.

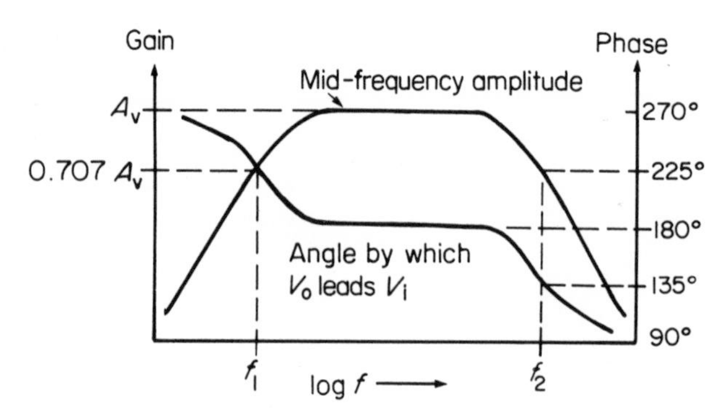

Figure 2.27

At the high-frequency end of the spectrum, a similar analysis will show that (in the coupling) v_o will lag v_i by some angle between 0° and 90°, and the overall phase shift will therefore lie between 180° and 90°. At the 3 dB points, where reactance is equal to resistance, the phase shift in the coupling will be ±45°, giving an overall shift of 180 ± 45°.

A plot of gain and phase shift curves for an R-C coupled amplifier therefore appears as shown in *Figure 2.27*.

Transformer coupling

An alternative method of coupling two stages together is shown in *Figure 2.28*. Here a transformer is used and this circuit has some advantages over *R-C* coupling. The d.c. resistance of the transformer windings is relatively small in comparison with the a.c. resistance, so

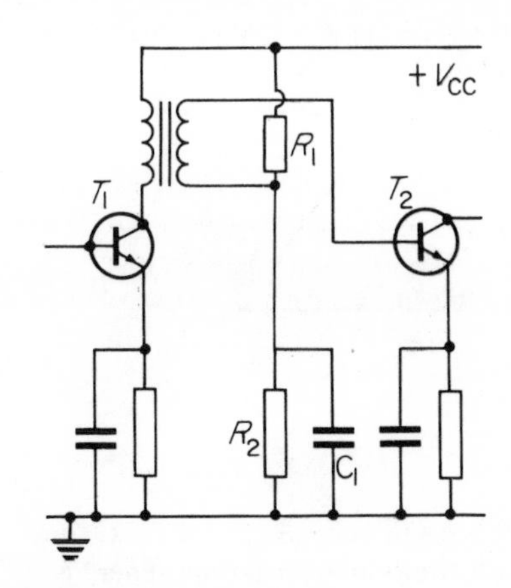

Figure 2.28

that while a large signal resistance is apparent in the collector circuit of T_1, the d.c. voltage drop is small. Hence there is only a small power loss in the resistance of the winding, a higher collector voltage is obtained for a given V_{cc} supply, and as a result the amplifier is more efficient. Further, the turns ratio of the transformer may be chosen so that the proper impedance matching is obtained between the transistors, and the maximum power gain (and hence current and voltage gains) obtained. Such impedance matching is not possible with *R-C* coupling. There are, however, disadvantages with transformer coupling, one being the greater cost.

Much of the basic theory associated with transformer coupling will be covered in Section 7. Here we will discuss briefly the effect of this coupling method on the response characteristics of the amplifier. In *Figure 2.28* the usual stabilised bias system is used as for *R-C* coupling, R_1 and R_2 providing base bias for T_2 by way of the low series d.c. resistance of the transformer secondary winding. Capacitor C_1 bypasses the signal to earth at the junction of R_1 and R_2, thus the whole of the input signal is delivered from the transformer secondary to the base of T_2 unaffected by the d.c. bias conditions.

Simple equivalent circuit diagrams for the transformer coupling at low, middle and high frequencies are given in *Figure 2.29(a), (b) and (c)* respectively. At low frequencies the parallel stray capacitances may be neglected, but the a.c. impedance (reactance) of the primary will be small. Hence the signal voltage developed across the primary will be small and the overall gain will be small. Over the mid-frequency range the stray capacitance will still have little effect, but the primary impedance will now be relatively high and increasing. Hence the gain will show a slowly rising characteristics.

At high frequencies the primary impedance will be very large but the stray capacitances and the capacitance of the winding itself will now add up to give an overall effective capacitance C_S in shunt with the

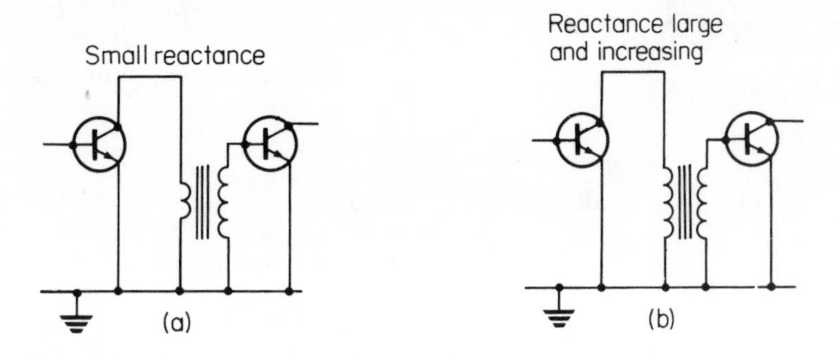

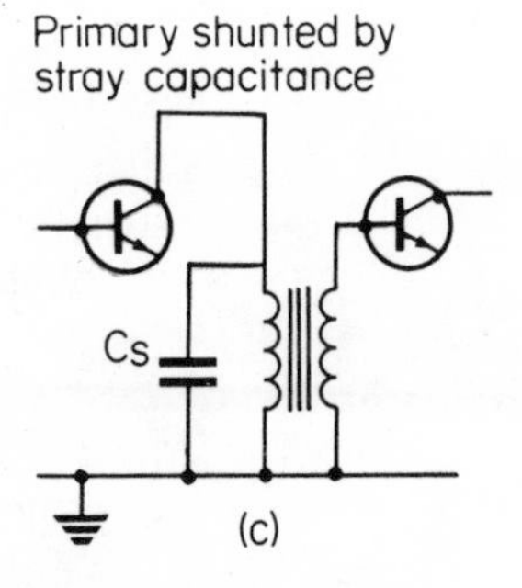

Figure 2.29

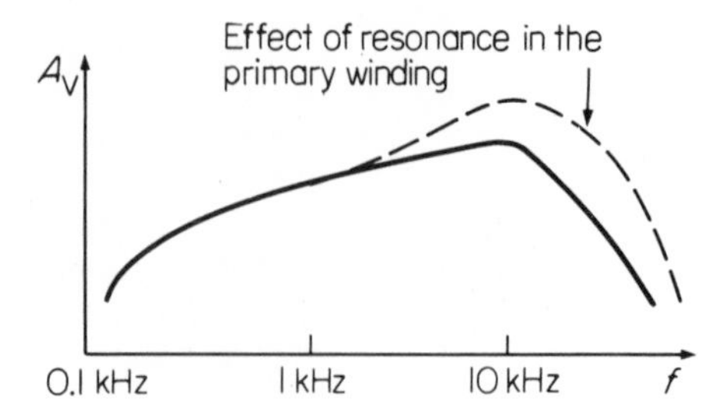

Figure 2.30

winding. The reactance of this capacitance will decrease as the frequency increases and progressively short-circuit the primary winding. There is often a point of resonance where C_S and the primary inductance act as a parallel tuned circuit of relatively high impedance, and at this point the response curve may exhibit an unusually large gain figure.

Figure 2.30 gives a general overall gain-frequency response curve of a transformer-coupled amplifier. The resonant rise at high frequencies can be reduced by increasing the winding resistances, so reducing the transformer Q; careful design can produce a response curve with the resonance point so adjusted that a usefully extended high frequency region is obtained.

The presence of direct current in the primary winding reduces the primary inductance and hence its reactance at a given frequency. For this reason, the circuit shown in *Figure 2.31* is sometimes used, the transformer primary being fed through a series capacitor C_C. This capacitor blocks out the d.c. component but allows the signal to pass on.

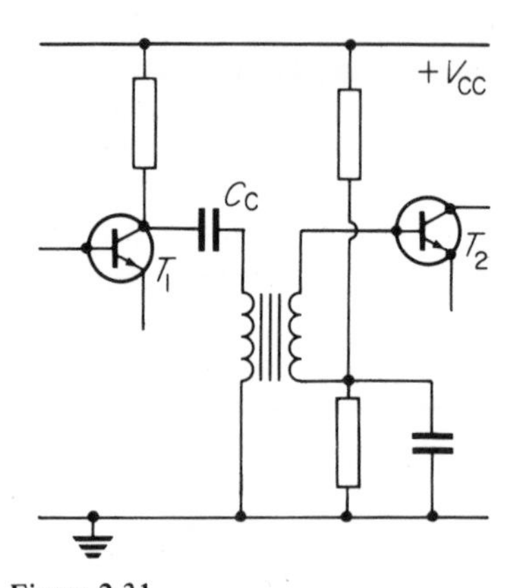

Figure 2.31

(14) Deduce the sort of gain-frequency response curve you would be likely to get from a circuit of the kind shown in *Figure 2.31* in comparison with the curve shown in *Figure 2.30.*

Direct coupling

The need often arises for an amplifier which will provide a high and stable gain at very low and zero frequencies. The low-frequency characteristics of R-C and transformer coupled amplifiers rule out their use in such applications as this and *direct* coupling between stages is then the only feasible answer. The problem with direct coupling is 'drift', the condition in which a gradual change occurs in the output voltage or current of the amplifier when the input signal itself is held at a constant level. In transistor amplifiers, drift depends upon:

(i) the collector supply voltage.
(ii) the temperature-dependent parameters of current gain, base-emitter voltage and leakage current.

The first of these variations can be eliminated by using a stabilised power supply, but the second is not always so easily dismissed. Leakage currents are a problem because of their extreme sensitivity to temperature. Such currents set up in the amplifier are amplfied along with the required signal currents, the amplifier being unable to distinguish between them. The problem is eased by the use of silicon transistors, and compensation is possible in the circuitry by employing temperature-sensitive resistors (thermistors) and diodes. More serious is the variation of base-emitter voltage with temperature. In any transistor V_{be} changes by about 2 mV for each °C change in temperature, decreasing as temperature increases. This may seem a negligible amount, but suppose the input transistor of a d.c. amplifier has a gain of, say, 20, then for every degree rise in temperature, the output will drift by 40 mV because the 2 mV variation in V_{be} will act as an input signal.

The performance of a d.c. amplifier is therefore described in terms of the *residual drift voltage*, which is that voltage which when applied to the input terminals of the first stage will produce the *same* change in the output current of the amplifier as is produced by the temperature dependent variations of the transistor parameters.

There are many circuit designs for direct-coupled amplifiers in which drift is reduced to very low figures. We can discuss here only one or two of the general arrangements used in the basic amplifier form.

SINGLE INPUT STAGES

Although d.c. amplifiers having a single-ended input stage are not common in quality designs, the principles involved are worth some general comments. It is possible to remove the coupling capacitor from an *R-C* amplifier, thereby placing the direct potential at the base of the second transistor at the same level as the collector of the first transistor. Such a circuit is shown in *Figure 2.32.* This is perfectly acceptable provided that the emitter and collector potentials of T_2 are adjusted to maintain the required operating voltages on this transistor.

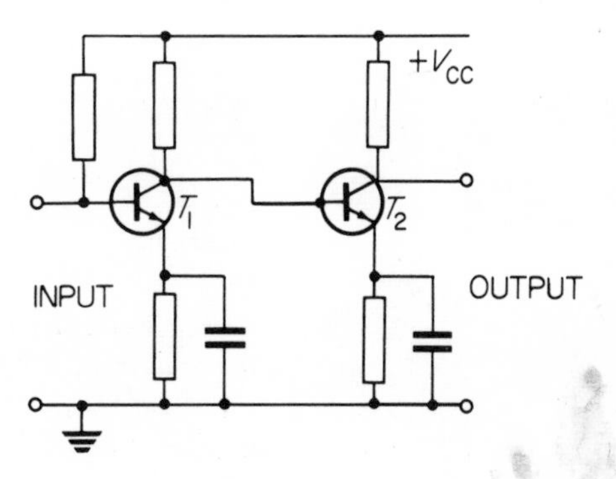

Figure 2.32

Obviously, in order to set T_2 to its proper operating point, something must be done to bring its emitter potential to a value slightly less than its base potential so that the correct base-emitter forward bias is obtained. When this has been done, the collector voltage of T_2 will be greater than the collector voltage of T_1, so for a number of such stages in tandem, the collector voltages will become even larger as successive stages are reached. The d.c. voltage at the output terminals will consequently be greater than the d.c. level at the input; so when the amplifier is inserted between a source and an output load, the d.c. level is shifted.

This is not important in a lot of cases, only voltage *changes* being of interest, but there is a problem in that the movement of the operating point along the load line eventually restricts the possible signal excursions of the later stages. The difficulty can be minimised by using an

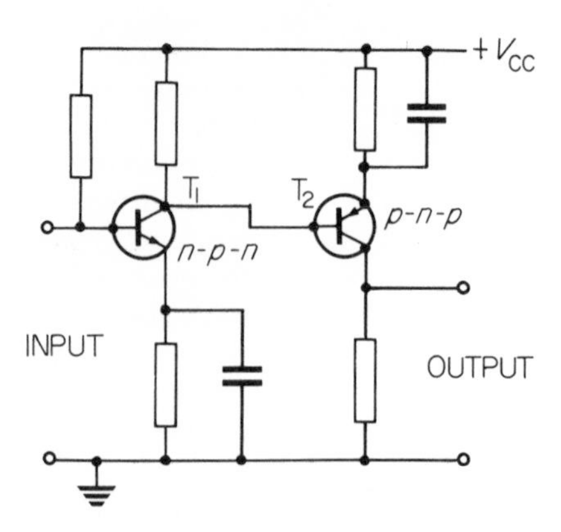

Figure 2.33

amplifier with alternate *n-p-n* and *p-n-p* transistors as shown in *Figure 2.33*.

By the proper choice of component values the emitter potential of T_2 can be made slightly less than the base potential, and the output, now taken across the collector load of T_2 can be made equal in mean d.c. level to the collector of T_1.

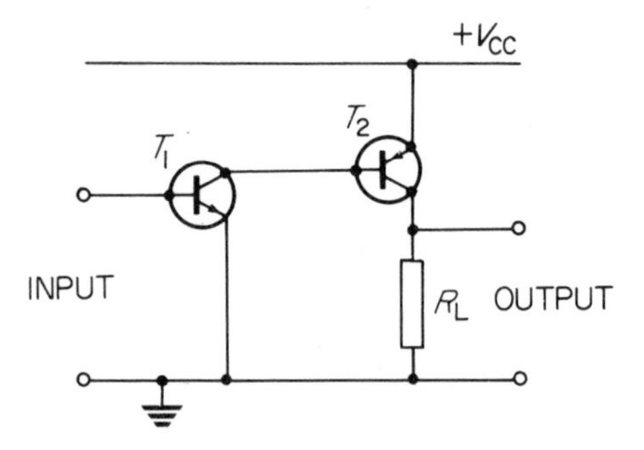

Figure 2.34

An alternative mixed transistor system is shown in *Figure 2.34.* Here the load resistor of T_1 is eliminated and the collector current of this transistor is the base current of T_2. The input voltage is assumed to provide forward bias for the transistor. If the signal source does not have a d.c. component which will provide the bias, a resistor must be connected between the base of T_1 and the V_{cc} line to provide the required base current.

The circuit of *Figure 2.33* can be improved by the substitution of a temperature-sensitive element for the collector load resistor. This could conveniently be a thermistor which has a negative temperature coefficient of resistance, that is, its resistance decreases as temperature increases. When the temperature increases, the collector current of T_1 will increase but the resistance of the thermistor will fall. Therefore the voltage drop across the thermistor and hence the emitter voltage of T_2 remain constant.

Simpie problems associated with direct coupled amplifiers are dealt with by Ohm's law and the basic current relationships in the transistors as the next worked example will illustrate.

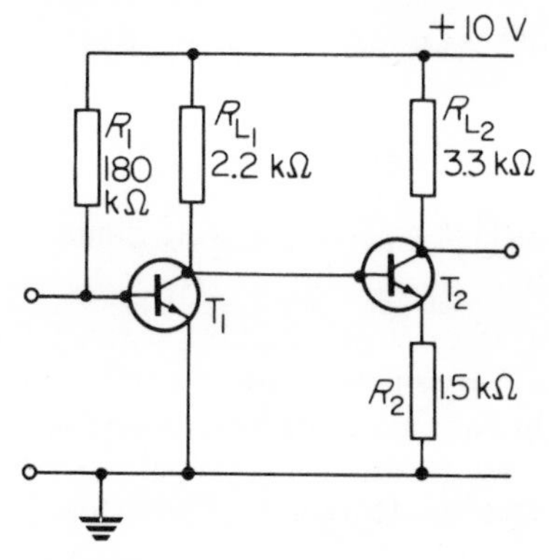

Figure 2.35

Example (15) In the amplifier of *Figure 2.35* the static current gains, α_1 and α_2 of T_1 and T_2 respectively are 65 and 30. If the voltage drop across R_{L1} is 8 V, calculate the collector and emitter voltages and currents of T_2. Assume that the base-emitter voltage of T_1 is negligible.

$$\text{Base current } I_{B1} \text{ of } T_1 \text{ is } \frac{10}{180 \times 10^3} \text{ A} = 55.5\ \mu\text{A}$$

Hence collector current I_{C1} of T_1

$$= \alpha_1 . I_{B1} = 65 \times 55.5\ \mu\text{A} = 3.6 \text{ mA}$$

Volts drop across $R_{L1} = 8$ V

$$\therefore \quad 8 = R(I_{C1} + I_{B2}) = 2.2(3.6 + I_{B2})$$

$$\therefore \quad I_{B2} = \frac{8 - 7.92}{2.2} = \frac{0.08}{2.2} = 0.036 \text{ mA}$$

Collector current I_{C2} of $T_2 = \alpha_2 \cdot I_{B2} = 0.036 \times 30$

$$= 1.08 \text{ mA}$$

$$T_2 \text{ emitter current} = \alpha_2 I_{C2} + I_{B2} = 1.08 + 0.036$$
$$= 1.116 \text{ mA}$$

$$T_2 \text{ emitter voltage} = 1.5 \times 1.116 = 1.674 \text{ V}$$

$$T_2 \text{ collector voltage} = V_{cc} - I_{C2} . R_{L2} = 10 - (1.08 \times 3.3)$$
$$= 6.44 \text{ V}$$

(16) Would there be any object in bypassing R_2 in *Figure 2.35* with a large value capacitor?

DARLINGTON CIRCUIT

A circuit configuration that is often found in d.c. amplifiers is the composite transistor or Darlington circuit shown in *Figure 2.36*. In this arrangement the emitter current I_{E1} of T_1 is the base current of T_2. If R_L is small, $I_{E1} = (\alpha_1 + 1)I_{B1}$ and $I_{C2} = \alpha_2 \cdot I_{B2}$ where α_1 and α_2 are the respective current gains of T_1 and T_2. Then the overall current gain

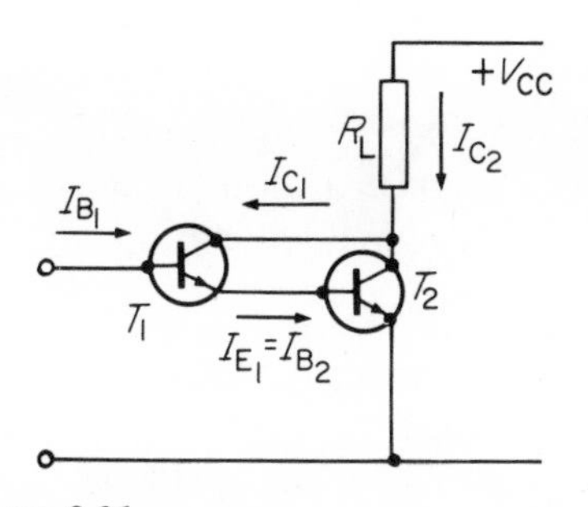

Figure 2.36

$$\frac{I_{C2}}{I_{B1}} = \frac{\alpha_2(\alpha_1 + 1)I_{E1}}{I_{B1}} = \alpha_2(\alpha_1 + 1)$$

since $I_{E1} = I_{B1}$.

Since α will be large in comparison with 1, the overall current gain is given very closely by the product $\alpha_1 \cdot \alpha_2$.

Darlington pairs and triples are available in a single package and provide very high current gains.

DIFFERENTIAL INPUTS

Because of the problem with drift, single-ended input stages are not often used in d.c. amplifiers. A circuit in common use is shown in *Figure 2.37* and is known as a *differential* input, so called because its output voltage is equal to the *difference* of two voltages applied to its separate inputs. Two transistors are employed in a balanced circuit

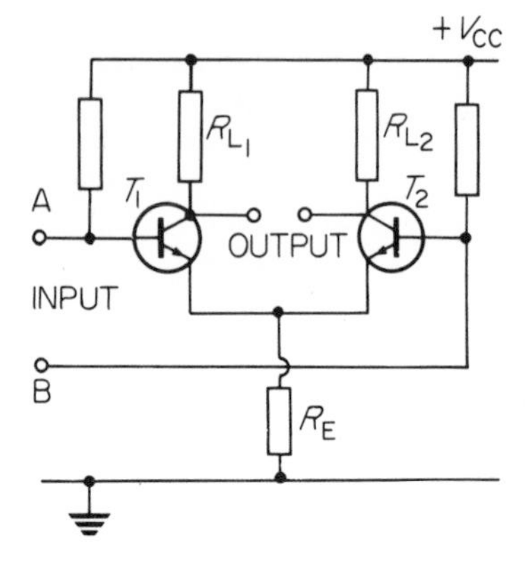

Figure 2.37

system although in certain applications the circuit may be unbalanced. For this circuit two input conditions can be considered:

(i) Differential-mode in which inputs of *opposite* polarity are applied to the inputs;
(ii) Common-mode in which identical in-phase voltages are simultaneously applied to the inputs.

For the differential mode input, a large output signal is generated. Suppose in the diagram that terminal A goes positive while B goes negative, T_1 will be driven into conduction and its collector potential will fall, while T_2 will be backed off and its collector potential will rise. The voltage appearing across the output terminals will therefore be large.

For a common mode signal input there will be theoretically no output as both collectors will shift by equal voltages in the *same* direction. Thus, any changes in the V_{be} of both transistors with temperature variation will act as common mode inputs and will be cancelled at the output. To help in this, the transistors must be mounted close together so that they are subjected to the same ambient conditions and must be matched in characteristics as closely as possible. The collector voltages of T_1 and T_2 may be fed to another differential pair for further amplification.

The circuit will amplify if only one input is provided and the other returned to earth through an appropriate resistance. This is an unbalanced circuit which is more usually known as an *emitter-coupled* amplifier.

A measure of the goodness of a differential amplifier is the *common mode rejection ratio* (CMRR) which is defined as

$$\frac{\text{Voltage gain for difference signals}}{\text{Voltage gain for common-mode signals}}$$

that is, it is the ratio of the output with equal input voltages of opposite phase (differential-mode) to the output obtained with equal in phase inputs. In a good circuit, this ratio will be very large.

PROBLEMS FOR SECTION 2

(17) 'A differential amplifier has a CMRR of 100'. Try to explain in a single sentence exactly what this statement implies.

(18) In a transistor, if I_C = 1.96 mA and I_e = 2.00 mA, what is I_b? If in addition I_{cbo} is 20 μA, what is α_E?

(19) Define: (a) amplitude distortion; (b) phase distortion; (c) frequency distortion as related to amplifier characteristics.

(20) Complete the following statements:

(a) The form of coupling least likely to interfere with the frequency response of an amplifier is . . . . . . coupling.

(b) In Class-A amplification, collector current flows for . . . of the input cycle.

(c) In resistance-capacitance coupling, the coupling capacitor affect the . . . . . frequency response of the amplifier.

(21) Express as decibel gains or losses, voltage ratios of 100, 40, 5 and 0.5 times. What power gains correspond to 20 dB, 7 dB, 6 dB and –30 dB? Why is it advantageous to use decibels in telecommunications engineering?

(22) An amplifier has resistive input and output impedances each of 75 Ω. When an input signal of 0.5 V is applied to the amplifier, a 10 V signal is developed at the output. Calculate (a) the voltage gain ratio; (b) the power gain ratio; (c) the power gain in dB.

(23) A low-frequency amplifier has a power gain of 56 dB. The input circuit is 600 Ω resistive and the output load is 10 Ω. What will be the current in the load when a 1 V r.m.s. signal is applied at the input?

(24) An amplifier has a response at 100 Hz which is 8.5 dB down on its response at 1 kHz. If the voltage gain at 100 Hz is 15, what is the voltage gain at 1 kHz?

(25) Show that voltage gain A_v can be expressed as $\alpha_E\ R_L/R_i$ where α_E is the current gain and R_i is the input resistance of the amplifier. Hence show that power gain $A_p = \alpha_E . A_v$.

(26) Explain carefully why a transistor must be biased by a method which prevents excessive shift of the d.c. working point. What effects are likely to follow if the d.c. stabilisation of the working point is insufficient? Give two methods of achieving bias and stabilisation in a common-emitter amplifier.

(27) A transistor with V_{be} = 0.6 V requires a base bias current of 75 μA. The V_{cc} supply is 6 V. What value of resistor should be connected between base and V_{cc} line to provide this bias?

(28) *Figure 2.38* shows a bias stabilisation circuit for a transistor. Assuming that the transistor has a constant base-emitter voltage of 0.7 V: (a) estimate the collector current for a static current gain (α_E) equal to 30; (b) what will the collector current change to if the transistor is replaced by one with α_E equal to 60? What conclusions do you draw from your solutions?

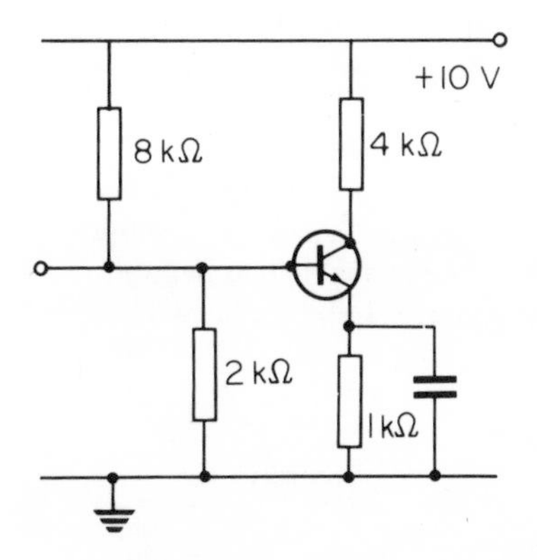

Figure 2.38

(29) The table shows the voltage gain of an amplifier whose input and output impedances are equal.

Frequency (kHz)	420	430	440	450	460	470	480
Voltage Gain	29.5	42	67	100	67	42	29.5

Plot the gain-frequency response curve on squared paper, gain being expressed in decibels. What is the 3 dB bandwidth of this amplifier, and what sort of amplifier is it?

(30) When the input voltage to a two-stage amplifier was maintained constant at 200 mV, the output voltages for various signal frequencies were as follows:

Frequency (kHz)	0.1	0.5	1.0	2.0	5.0	10	15
Output voltage (V)	1.42	5.6	7.1	6.32	2.8	0.8	0.36

Plot the gain-frequency characteristic of this amplifier and estimate its 3 dB bandwidth. What is the probable form of coupling used between the stages of this amplifier?

(31) The following table gives the output characteristic data of a transistor in common-emitter configuration.

V_{ce} (V)	I_C (mA) I_b = 40 μA	60 μA	80 μA
2	2.9	4.4	5.9
6	3.7	5.4	7.0
10	4.5	6.4	8.2

Sketch the output characteristics for the stated values of base current. Using the curves, estimate: (a) the output resistance of the transistor for I_b = 60 μA, (b) the current gain for V_{ce} = 7 V.

(32) A transistor has the following characteristics which may be taken to be linear between the given values of collector voltage:

V_{ce}	I_C (mA) I_b = 20 μA	40 μA	60 μA	80 μA
1	0.9	1.9	3.0	4.0
6	1.28	2.41	3.66	4.8

The transistor is used as a common-emitter amplifier with a collector load of 1.5 kΩ and a V_{cc} supply of 8 V. By choosing a suitable operating point, estimate the voltage, current and power gain of the amplifier when an input signal current of 60 μA peak-to-peak is applied at the base terminal. The input resistance of the transistor may be taken as 1.2 kΩ.

(33) Find the frequency at which a resistance of 10 kΩ shunted by 500 pF has fallen 3 dB below its mid-frequency value.

(34) Sketch a simple equivalent circuit for *Figure 2.39* for low, middle and high frequencies. Estimate the probable bandwidth of this amplifier, assuming that only the inter-stage coupling will affect this.

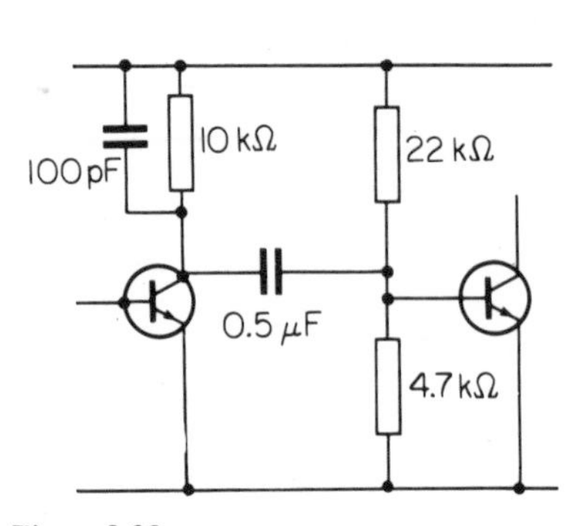

Figure 2.39

(35) Explain what is meant by the term 'drift' as applied to a directly coupled transistor amplifer and state the factors upon which it depends. Define common-mode rejection ratio for a differential amplifier.

(36) The voltage gain of an amplifier is 20 when terminated by a load resistor R ohms. This resistor is replaced by a 10 dB attenuator presenting the same resistance R ohms and its output is then terminated by the load R. Calculate (a) the required voltage input to the amplifier to give 50 mV across the load resistor R, (b) the power dissipated into the attenuator when R = 1000 ohms.

(37) A d.c. amplifier has a voltage gain of 40 dB. When the ambient temperature changes from 20°C to 30°C the output changes by 0.1 V. What is the magnitude of drift expressed as μV per degree referred to the input?

3 The field effect transistor

Aims: At the end of this Unit section you should be able to:
Compare the properties of the field-effect transistor with bipolar transistors.
Describe the basic construction of field-effect transistors and explain their principles of operation.
Explain the difference between depletion and enhancement modes.
Determine the form of the output and transfer characteristics and derive the basic operating parameters from these.
Calculate the stage gain of a common-source amplifier with a resistive drain load.
Explain the effect on frequency response and general performance of using loads other than resistive loads.
Understand the operation of a field-effect transistor as a switch.

We turn now to the field-effect transistor (FET). This is a device which exhibits certain characteristics that are markedly superior to those of bipolar junction transistors and which operates on the principle that the effective cross-sectional area and hence the resistance of a conducting rod of semiconductor material can be controlled by the magnitude of a *voltage* applied at the input terminals.

The FET operates upon a completely different principle from bipolar transistors. In these, the junction has been in series with the main current path from emitter to collector, and the operation of the transistor has depended upon the injection of majority carriers from the emitter into the base region. There is no such injection in FETs which depend only upon one effective *p-n* junction and only one type of charge carrier. For this reason FET's are known as *unipolar* transistors.

As an amplifier the FET has a very high input impedance comparable with that of a thermionic valve, generates less noise than the ordinary transistor, has high power gain and a good high frequency performance. In addition it has a large signal handling capability, voltage swings at the input being measured in volts. At the best, base voltage swings on bipolar transistors is measured in fractions of a volt.

There are several forms of FET and these are discussed below.

THE JUNCTION GATE FET

In its simplest form, the junction gate FET (or JUGFET) is constructed as shown in *Figure 3.1*. Here a length of semiconductor material, which may be either *p*- or *n*-type crystal, has ohmic (non-rectifying) contacts made at each end. The length of semiconductor is known as the *channel* and the end connections form the *source* and the *drain*. We shall assume throughout this discussion that the channel is *n*-type material as this form of construction is the most commonly used in practical designs.

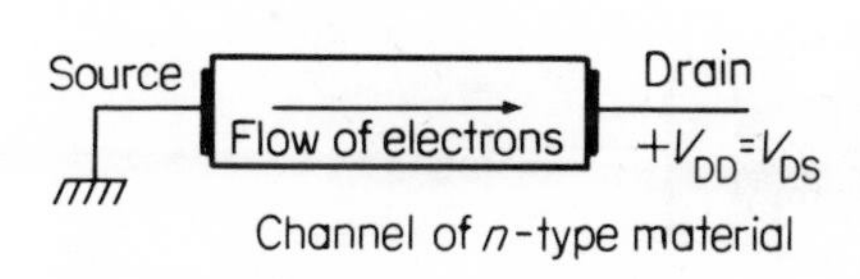

Figure 3.1

With no voltages applied to the end connections, the resistance of the channel $R = \rho l/A$ where ρ is the resistivity of the material and l and A are the length and cross-sectional area of the channel respectively. For example, if ρ = 5 ohm-cm, l = 0.1 cm and A = 0.001 cm^2, the channel resistance is 500 Ω. If the source end of the channel is effectively earthed and the drain end is taken to a positive potential, a current will flow along the channel (conventionally) from source to drain. This is drain current I_D. Clearly, if the effective cross-sectional area of the channel can be varied by some means, its resistance and hence the drain current will be brought under external control.

A means of varying the resistance of the channel is shown in *Figure 3.2(a)*. Two *p*-type regions, known as *gates*, are positioned one on each side of the channel. If these two gates are connected to each other and to the source terminal, they form reverse-biased diode *p-n* junctions with the channel crystal and depletion layers will be established as shown in *Figure 3.2(b)*. As the channel has finite conductivity, there will be an approximately linear fall in potential along the channel from the positive drain terminal to the zero (earth) potential at the source terminal, hence the contours of the depletion layers will take the form shown in diagram (b), being widest at the drain end because the effective reverse bias voltage between channel and gate is greatest at that end. The flow of electrons from source to drain is now restricted to the wedge-shaped path shown which represents a channel of reduced cross-sectional area compared with the normal condition of diagram (a).

The effective area of the channel is clearly dependent upon the drain-source potential, V_{DS}, because if this potential is increased, the depletion regions will grow and eventually meet; the channel conduction area is then reduced to zero at a point towards the drain end of the channel as *Figure 3.2(c)* depicts. The channel is now said to be *pinched-*

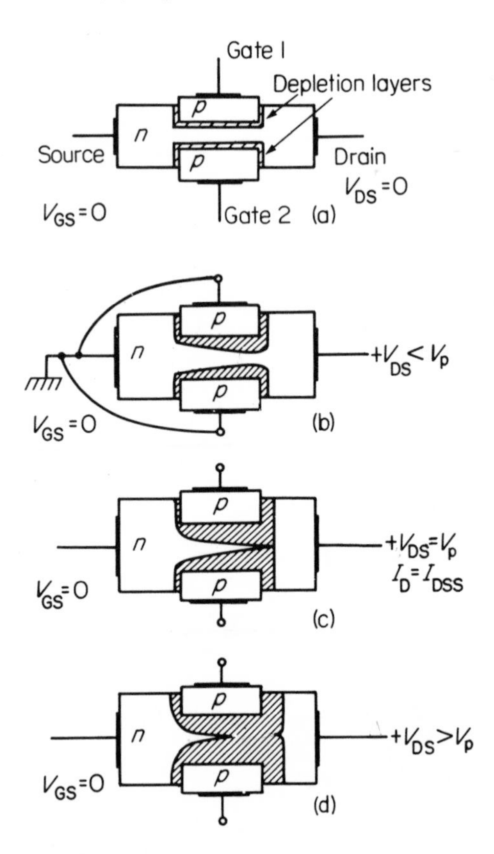

Figure 3.2

off, and the value of V_{DS} at which this occurs is known as the ***pinch-off*** voltage, V_p.

It is important to take note of the fact that the drain current does not cease when the drain voltage reaches pinch-off because a voltage equal to V_p still exists between the pinch-off point and the source, and the electric field along the channel causes the carriers (electrons) in the channel to drift from source to drain. However, because of the high effective resistance of the channel, the drain current does become substantially independent of the drain voltage.

As V_{DS} is increased beyond V_p the depletion layer thicken between the gate and the drain as shown in *Figure 3.2(d)* and the additional drain voltage is effectively absorbed by the increased field in the wider pinched-off region. The electric field between the original pinch-off point and the source remains substantially unaffected, hence the channel current and so the drain current remains constant. Electrons which arrive at the pinch-off point find themselves faced by a positive potential and are swept through the depletion layer region in exactly the same way as electrons are swept from base to collector in a bipolar *n-p-n* transistor.

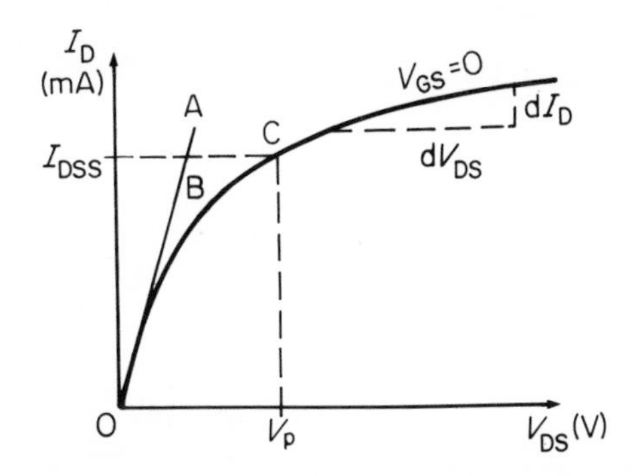

Figure 3.3

It is now possible to represent the relationship between the drain current I_D, and the drain-source voltage V_{DS} in graphical form. In *Figure 3.3* the line OA represents the behaviour of the channel acting simply as a semiconductor resistor; the drain current I_D would follow this line for increases in V_{DS} if the resistance of the channel was constant, unaffected that is, by the effect of the increased drain voltage upon the width of the depletion regions. At point B the action of the depletion layers begins to take effect and there is a departure from the linear characteristic. At point C the applied V_{DS} reaches V_p and pinch-off occurs. For values of V_{DS} greater than V_p the channel remains pinched off and I_D becomes virtually constant and independent of V_{DS}. This characteristic curve is for the particular case when the gate-source potential is zero and is so indicated on the diagram. The drain current which flows when $V_{GS} = 0$ and $V_{DS} = V_p$ is indicated as I_{DSS}, representing the saturated (constant) drain current with the input short-circuited i.e. gates connected to source.

Suppose now that the gates are negatively biased relative to the source instead of being at zero potential. The depletion regions will clearly be thicker for a given value of V_{DS} than they were when there was no such negative bias on the gates. As a result, pinch-off and saturation drain current will occur at lower values of V_{DS}, and when

V_{GS} is sufficiently negative, the electric field between the source and the original pinch-off point will be eliminated and drain current will cease to flow. This situation is illustrated in *Figure 3.4(a)*, and this diagram should be compared to the pinch-off condition of $V_{GS} = 0$ in *Figure 3.2(c)*.

In *n*-channel FETs the gate must be negative with respect to the source to achieve pinch-off. The reverse polarity is necessary for a p-channel device. The sign of V_p is therefore as follows:

n-channel $V_p < 0$
p-channel $V_p > 0$

In operation as an amplifier the FET is biased so that $V_{DS} > V_p$ and $V_{GS} < V_p$. This condition is illustrated in *Figure 3.4(b)*.

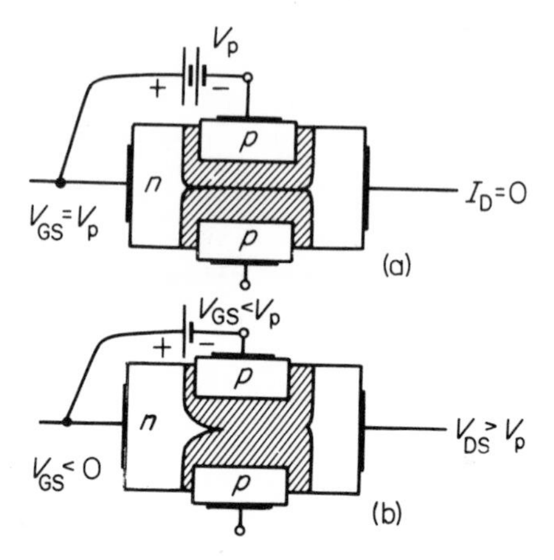

Figure 3.4

Characteristic curves

There are two characteristic curves of interest to use here, the *output* or *drain characteristic* and the *transfer* or *mutual characteristic.*

We have already touched on the form of the output characteristic in discussing *Figure 3.3* and we can now extend that diagram to include the effect of an increasing negative bias on the gate of the transistor. This has been done in *Figure 3.5* and a family of curves relating I_D to V_{DS} for different values of V_{GS} has been obtained. Notice from this diagram that it is possible to operate the junction FET with a small

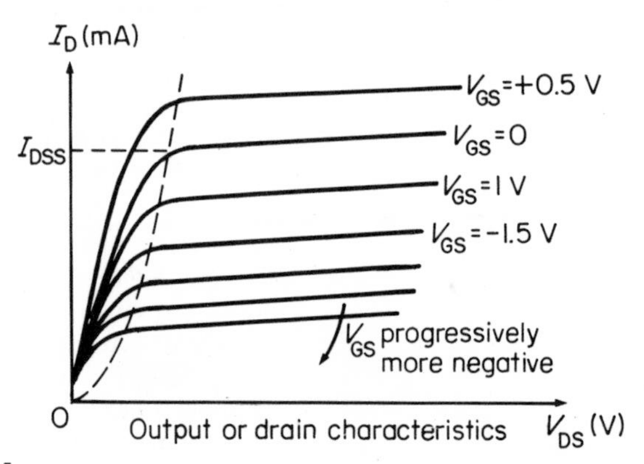

Figure 3.5

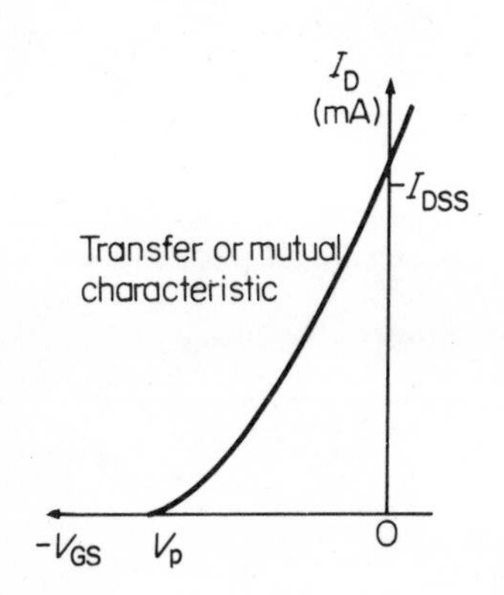

Figure 3.6

positive gate bias. The superficial similarity of these curves to the output characteristics of bipolar transistors should be noted.

The broken line in the diagram represents the locus of a point passing through the respective pinch-off locations on each separate curve. To the left of this line the FET behaves as a variable resistor and this region is known as the *ohmic* region. To the right of the line the FET behaves as a constant current generator and exhibits a very high output resistance, just as the bipolar transistor does. As mentioned, the FET is normally operated in this saturation region where values of V_{DS} exceed V_p.

The transfer or mutual characteristic relates the dependence of drain current I_D upon the gate-source voltage V_{GS} and a typical characteristic is shown in *Figure 3.6.* There is actually a family of these curves, each relating to a particular value of V_{DS}, but as I_D is practically constant beyond the pinch-off point, the curves are almost coincident with one another. To avoid confusion, only a single curve is shown in the figure. Notice that the particular value of I_D for which V_{GS} is zero is I_{DSS}, in accord with the output characteristics of *Figure 3.3* and *Figure 3.4.*

The pinch-off voltage V_p appears on this characteristic as well as on the output characteristics because the channel can be pinched-off by applying V_p between gate and source in the reverse direction. The depletion layers then meet along the whole length of the channel and $I_D = 0$. Strictly, a small leakage current continues to flow along the channel which varies with temperature; this temperature dependence of V_p is due entirely to the variation of barrier potential at the *p-n* junctions, the same effect that causes variation of V_{BE} with temperature in junction transistors. We shall return to problems of temperature effects a little later on.

FET parameters

In order to analyse the amplifying properties of the junction FET, we borrow two parameters from thermionic valve theory. These parameters are mutual conductance and drain slope resistance.

1. Mutual conductance g_{fs} (or g_m) is defined

$$g_{fs} = \frac{\text{the small change in } I_D}{\text{the small change in } V_{GS}} \quad \text{for a constant } V_{DS}$$

$$= \left.\frac{dI_D}{dV_{GS}}\right|_{V_{DS}=k}$$

This parameter, relating the dependence of drain current I_D to the gate-source voltage V_{GS}, represents the gradient of the transfer characteristic of the FET.

As g_{fs} represents the ratio of a current to a voltage, its unit of measurement will be the Siemen. For a junction FET, g_{fs} will range typically from 10^{-2} to 10^{-3} S.

The equation of the characteristic is complicated, but to a good approximation I_D varies as the square of V_{GS} so that the curve is closely parabolic. It can be shown experimentally that the relationship between I_D and V_{GS} can be expressed in the form

$$I_D \simeq I_{DSS}\left[1 - \frac{V_{GS}}{V_p}\right]^2 \tag{3.1}$$

Notice from this that when V_{GS} is zero, $I_D = I_{DSS}$ as it should.

Example (1) Using the equation just stated, derive an expression for the mutual conductance of a FET. What is the mutual conductance for the case where $V_{GS} = 0$?

Since $g_{fs} = dI_D/dV_{GS}$, the mutual conductance can be obtained by differentiating equation (3.1), noting that V_p is constant. Then

$$g_{fs} = 2I_{DSS}\left[1 - \frac{V_{GS}}{V_p}\right]\left[-\frac{1}{V_p}\right]$$

$$= \frac{-2I_{DSS}}{V_p}\left[1 - \frac{V_{GS}}{V_p}\right] \text{ siemens}$$

The value of g_{fs} obtained at zero bias, or when $V_{GS} = 0$, is denoted by g_{fso} and this follows at once from the previous result:

$$g_{fso} = -\frac{2I_{DSS}}{V_p} \text{ siemens}$$

Notice that g_{fs} or g_{fso} will *always* be positive since either V_p *or* I will be negative, whether the transistor is *n*- or *p*-channel. The gradient of the mutual characteristic is clearly positive as a glance will show.

(2) Can you see why the expression for g_{fso} above would be useful in a practical method of measuring the g_{fs} of a transistor; *Hint*: substitute g_{fso} back into the equation for g_{fs}, and then think about how you would measure the various parameters.

Example (3). The output characteristics for the 2N5457 FET are shown in *Figure 3.7.* Using these curves, derive the mutual characteristic for this transistor for $V_{DS} = 10$ V. Hence estimate the mutual conductance of the 2N5457 at a gate bias of –0.5 V.

This example will illustrate how the mutual characteristic can be obtained from the output characteristics. From *Figure 3.7* we notice that V_{GS} ranges from zero to –1.5 V while I_D ranges from zero to just over 3 mA. Accordingly we draw the axes for the required mutual characteristic to cover the same ranges. These are shown in *Figure 3.8.*

Referring now to the $V_{DS} = 10$ V point on the output horizontal axis, we make a note of the I_D values where a line erected vertically from this point cuts the V curves; these are the points A, B, C, etc. These I_D values are then plotted on the mutual characteristic axes against the corresponding values of V_{GS}. Connecting the points so obtained, the required characteristic curve appears as the diagram of *Figure 3.8* shows. To find g_{fs} at the point where $V_{GS} = -0.5$ V (point P), we draw a tangent to the curve, and then

$$g_{fs} = \frac{dI_D}{dV_{GS}} = \frac{2.3 \times 10^{-3}}{1.1} = 2.1 \times 10^{-3}\ \text{S}$$

$$= 2.1\ \text{mS}$$

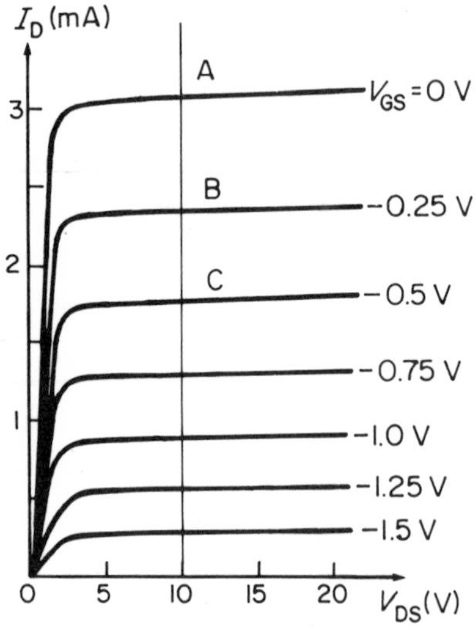

Figure 3.7

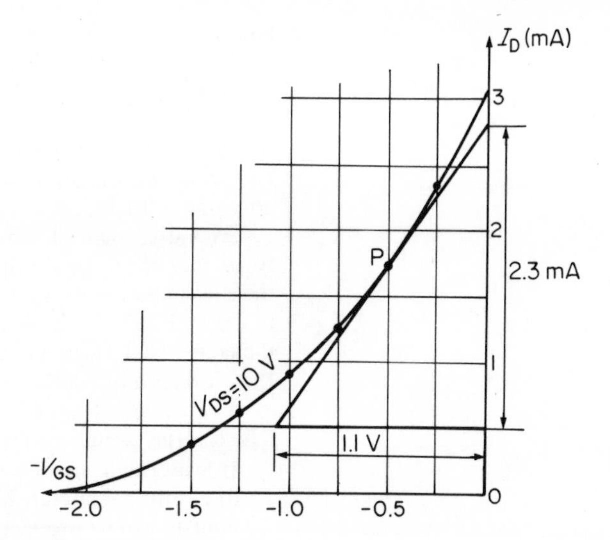

Figure 3.8

As in valve theory, g_{fs} is sometimes expressed in mA per volt; the above solution might equally well be written as 2.1 mA/V. This then tells us that the drain current changes by 2.1 mA when the gate voltage changes by 1 V on a bias of –0.5 V.

We come now to the second of the FET parameters:
2. Drain slope resistance r_d (or r_{ds}) is defined

$$r_d = \frac{\text{the small change in } V_{DS}}{\text{the small change in } I_D} \text{ for a constant } V_{GS}$$

$$= \left.\frac{dV_{DS}}{dI_D}\right|_{V_{GS} = k}$$

This parameter, relating the dependence of drain current I_D upon the drain-source voltage V_{DS} represents the reciprocal of the gradient of the output characteristic in the saturation region. *Figure 3.3* shows the meaning of drain slope resistance. You should observe that the gradients of the drain characteristics are, like the output curves of a bipolar transistor, very flat in the saturation region and hence the output resistance is very high.

THE JUGFET AS AN AMPLIFIER

Except for the rather different biasing required, the junction FET described above can be used in any of the transistor circuits already discussed. The drain, gate and source are loosely equivalent to the collector, base and emitter of the bipolar transistor, or to the anode, grid and cathode of a thermionic valve. There are, however, two very important differences between the FET and the ordinary transistor which must be emphasised.

First, the control of current flow through the FET is by way of V_{GS}, that is, a *voltage* control and not a current control. Because the input junctions are reverse-biased, the gate leakage current is negligible and the input resistance of the FET is correspondingly very high, of the order of 10^3 to 10^6 megohms. Although the input resistance decreases rapidly when the junctions become forward biased, the input resistance remains relatively high (a megohm or more) in a silicon device provided the forward bias does not exceed some 0.5 V at ordinary temperatures. The FET can therefore be operated as a small signal amplifier with $V_{GS} = 0$.

Secondly, the FET is a majority carrier device; in the *n*-channel form only *electrons* are the carriers drifting from source to drain. In the *p*-channel form only *holes* are the carriers drifting from source to drain, V_{DS} now being of reversed polarity. In both respects, the *n*-channel FET resembles the thermionic valve much more closely than it does a bipolar transistor.

We shall be interested here only in the *common-source* configuration which is the circuit equivalent of the common-emitter connection.

It is useful at this stage to relate the two FET parameters g_{fs} and r_d to a third parameter. We have noted that as far as I_D is concerned, its control can be brought about either by varying V_{DS} (the output characteristic) *or* V_{GS} (the mutual characteristic). The gate voltage V_{GS} exerts a much greater influence on I_D than does the drain voltage V_{DS} for a given variation of potential. Suppose a small increase in V_{DS} causes an increase in I_D, and that I_D can then be restored to its original value by a small negative change in V_{GS}. The ratio of these two changes in the drain and gate voltages which produce *the same change* in I_D is called the *amplification factor* of the FET and is symbolized μ. So

$$\mu = \frac{\text{the small change in } V_{DS}}{\text{the small change in } V_{GS} \text{ producing the same change in } I_D}$$

$$= \left. \frac{dV_{DS}}{dV_{GS}} \right|_{I_D = k}$$

Since μ is the ratio of two voltages, it is simply expressed as a number. Amplification factor is related to g_{fs} and r_d because the product

$$g_{fs} \times r_d = \frac{dI_D}{dV_{GS}} \times \frac{dV_{DS}}{dI_D} = \frac{dV_{DS}}{dV_{GS}} = \mu$$

Hence

$$\mu = g_{fs} . r_d \tag{3.2}$$

The circuit, and the circuit symbol, for an *n*-channel FET common-source amplifier is shown in *Figure 3.9.* The input signal V_i is applied between gate and earth, and the drain circuit contains a load resistor R_L across which the output voltage V_o is developed by the flow of drain current. A resistor R_S is included in the source lead and the drain current also flows through this. A voltage equal to $R_S I_D$ is therefore developed across R_S and the source terminal is raised by this potential above the earth line. As the gate is connected to earth through resistor R_G the source is effectively positive with respect to the gate, or, what is the same thing, the gate is *biased negatively* with respect to the source. There is no d.c. voltage developed across R_G because the gate current is negligible; at the same time R_G presents a high impedance

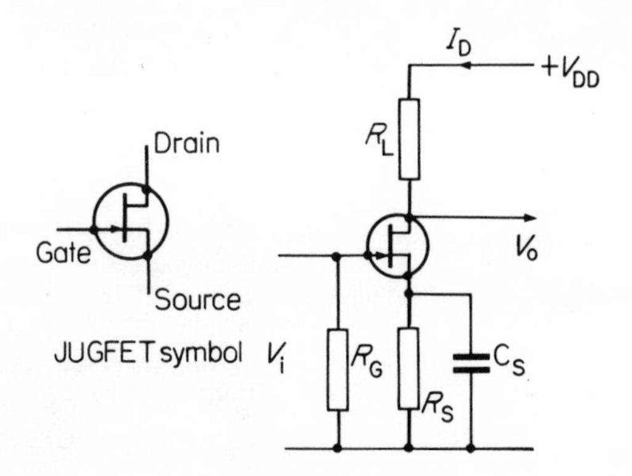

Figure 3.9

to the a.c. input signal. Notice that the drain supply voltage is designated V_{DD}.

Suppose the gate voltage to move positively by a small amount due to the input signal; then I_D will increase and there will be an increased voltage drop across R_L. Hence the drain voltage V_D will fall. Like the bipolar common-emitter amplifier, the common-source FET amplifier introduces a 180° voltage phase change. This negative change in drain voltage in turn causes a further change in drain current, *separate* from but *simultaneous* with the change in drain current due to the original change in gate voltage. The *total* change in I_D is therefore the sum of two simultaneous changes in V_i and V_D. Let a change dV_i in input voltage cause a change dI_D in drain current. Then by definition of g_{fs}

$$dI_D = g_{fs}.dV_i$$

This change in I_D will in turn produce a change dV_D in drain voltage, so by definition of r_d

$$dI_D = \frac{1}{r_d}.dV_D$$

But $dV_D = -I_D.R_L$, the negative sign indicating that dV_D falls as I_D increases. The total change in I_D is therefore

$$dI_D = g_{fs}.dV_i - \frac{1}{r_d}.dI_D.R_L$$

Rearranging:

$$dI_D = \frac{g_{fs}.r_d}{r_d + R_L} dV_i = \frac{\mu}{r_d + R_L}.dV_i$$

But the output voltage change across R_L is $dV_o = -dI_D.R_L$ hence

$$dV_o = -\frac{\mu R}{r_d + R_L}.dV_i$$

$$\therefore \frac{dV_o}{dV_i} = A_v = -\frac{\mu R_L}{r_d + R_L} \tag{3.3}$$

which is an expression for the voltage gain of the FET common-source amplifier.

If we assume that r_d is very much *greater* than R_L (which is true in many amplifiers), the expression for voltage gain approximates to

$$A_v \simeq -\frac{\mu R_L}{r_d} \simeq -g_m.R_L$$

Since the signal input current to a FET is negligible, the current gain is very high but of little importance, and we shall not pursue it further.

Example (4) In a common-source amplifier, the parameters for the FET are $g_{fs} = 2.5$ mS, $r_d = 100$ kΩ. If the load resistor is 22 kΩ, calculate the voltage gain of the amplifier.

From equation (3.3) $A_v = -\frac{\mu R_L}{r_d + R_L}$

But $\mu = g_{fs} \times r_d = 2.5 \times 10^{-3} \times 100 \times 10^3$

$= 250$

$\therefore \quad A_v = -\frac{250 \times 22000}{122000}$

$= -45$

It is now possible to derive an equivalent circuit for the FET amplifier, for equation (3.3) connects voltage output V_o, a constant voltage source of e.m.f. $\mu. V_i$ volts, and a resistance made up of r_d and R_L in series. Rewriting the equation gives us

$$V_O = -\mu. V_i \times \frac{R_L}{r_d + R_L}$$

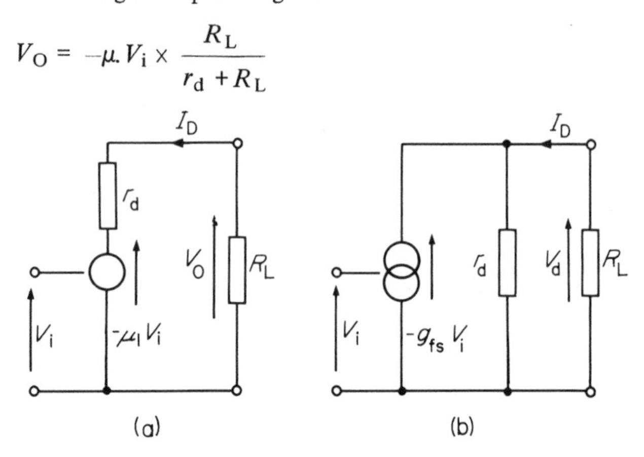

Figure 3.10

and this can be represented by the circuit shown in *Figure 3.10(a)*. The FET is replaced by a *constant-voltage* generator whose e.m.f. is $-\mu.V_i$ and this sends current through a resistance made up of r_d and R_L in series, the output voltage V_o being that fraction of $-\mu. V_i$ developed across R_L.

As the FET behaves as a constant-current generator, however, it is often better to use a constant-current form of equivalent circuit and this can be done by expressing equation (3.3) in the form

$$V_o = -g_{fs}.V_i \frac{r_d.R_L}{r_d + R_L}$$

where μ has been replaced by the product $r_d.g_{fs}$. This now expresses V_o in terms of a constant current $g_{fs}V_i$ flowing in a circuit made up of r_d and R_L in parallel. This circuit replaces the FET by a generator which provides a *constant-current* output feeding a parallel arrangement of r_d and R_L. This circuit is shown in *Figure 3.10(b)*. We have assumed for both forms of equivalent circuit that the input resistance of the FET is infinitely high and that the device capacitances are negligible. That is, we are considering a low-frequency equivalent circuit representation.

(5) A voltage gain of 20 is required from a FET having g_{fs} = 2.0 mS and r_d = 50 kΩ. What should be the value of the drain load resistor to provide this gain?

Example (6) In the amplifier of *Figure 3.9*, V_p = –2 V and I_{DSS} = 2.0 mA. The circuit is to be biased so that I_D = 1.2 mA, the drain supply V_{DD} being 20 V. Estimate (a) V_{GS}; (b) mutual conductance g_{fs}; (c) source bias resistor R_S; (d) the required value of R_L to give a voltage gain of 15.

(a) From our basic equation

$$I_D \simeq I_{DSS} \left[1 - \frac{V_{GS}}{V_p}\right]^2$$

we have by rearrangement

$$V_{GS} \simeq V_p \left\{1 - \left[\frac{I_D}{I_{DSS}}\right]^{\frac{1}{2}}\right\}$$

Check this for yourself before going on. Now inserting the given values we obtain

$$V_{GS} \simeq 2\left\{1 - \left[\frac{1.2}{2}\right]^{\frac{1}{2}}\right\} \simeq -0.45 \text{ V}$$

(b) Here we can use the equation

$$g_{fs} = -\frac{2I_{DSS}}{V_p}\left[1 - \frac{V_{GS}}{V_p}\right]$$

and inserting values

$$g_{fs} = -\frac{2 \times 2}{-2}\left[1 - \frac{0.45}{2}\right] = 1.55 \text{ mS}$$

(c) By Ohm's law

$$R_S = \frac{-V_{GS}}{I_D} = \frac{0.45 \times 10^3}{1.2}$$

$$= 375\ \Omega$$

(d) We have to use the approximate expression for voltage gain here as we do not know the values of r_d and μ. Assuming that r_d is very large compared with R_L, we have

$$A_v = -g_{fs}.R_L$$

or

$$-15 = -1.55 \times 10^{-3} \times R_L$$

from which

$$R_L = 9.7\ \text{k}\Omega$$

A 10 kΩ resistor would be used here.

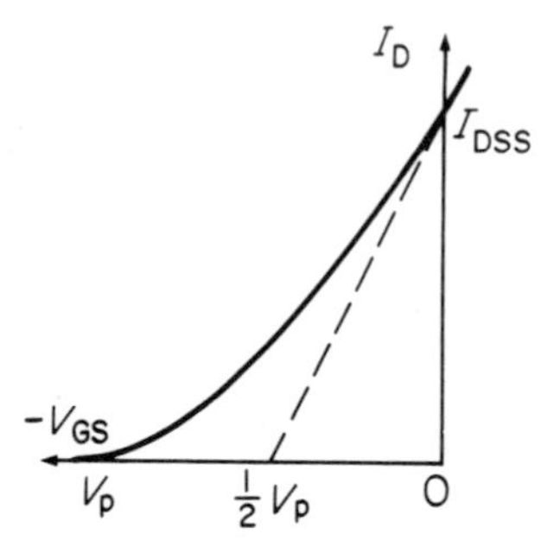

Figure 3.11

(7) *Figure 3.11* shows a mutual characteristic curve for a FET. Prove that if the gradient of the curve at the point where V_{GS} is 0 is continued as the diagram shows, it will cut the horizontal axis at the point $V_p/2$.

(8) The following values are taken from the linear portions of the static characteristics of a FET:

V_{DS}	15	15	10	volts
I_D	13.5	10.5	12.7	mA
V_{GS}	−0.5	−1.0	−0.5	volts

Calculate the parameters r_d, g_{fs} and μ for this FET.

LOAD LINE ANALYSIS

When a load resistor is placed in the drain output circuit to provide a signal voltage output, a load line can be drawn across the output characteristics as it was for the bipolar transistor amplifier. *Figure 3.12* shows a typical graph of this sort, the gradient of the load line depending upon whether an a.c. or a d.c. load condition is being con-

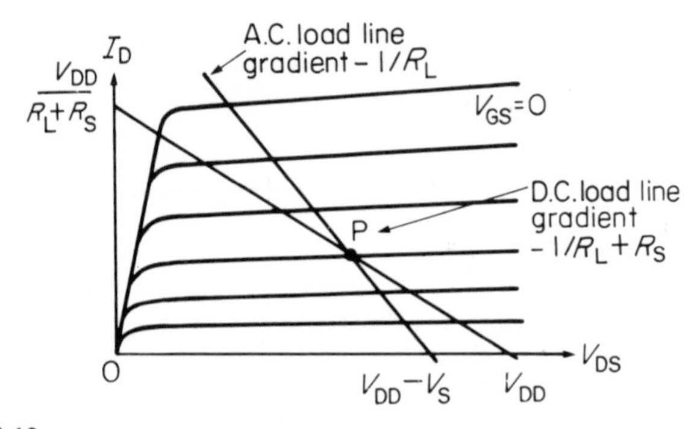

Figure 3.12

sidered. The d.c. load line has a gradient given by $-1/(R_L + R_S)$ and cuts the horizontal axis at V_{DD}, since for $I_D = 0$ the drain voltage must rise to the applied voltage. The other end of the line assumes that the volts drop across the FET is zero at saturation, hence the drain current will be $V_{DD}/(R_L + R_S)$, R_L and R_S being effectively in series across V_{DD}.

The a.c. load line has a gradient given by $-1/R_L$ since R_S is effectively short-circuited by capacitor C_S (see *Figure 3.9*) at signal frequencies, and so cuts the horizontal axis at $(V_{DD} - V_S)$ For Class A operation both load lines must pass through the operating point P which lies on the V_{GS} characteristic corresponding to the d.c. gate-to-source voltage developed across R_S. The procedure for determining the load lines is, in other words, similar to that already described for bipolar transistors and the method of using them to obtain information on the amplifier performance is also identical. The procedure can therefore be summarized here as follows:

(a) The input signal excursions applied to the gate terminal of the amplifier centred on the operating point P must be such that the peaks do not intrude beyond the knee of the $V_{GS} = 0$ characteristic (saturation) or below the horizontal axis limit (cut-off).

(b) The d.c. output drain voltage and drain current at quiescence determine the position of P and this must be chosen with regard to the conditions outlined in (a) above. P will normally be positioned so that the signal swings over characteristic curves which are linear and equally spaced.

THE INSULATED GATE FET

It is the high input resistance of the junction FET which makes it a particularly attractive device in many applications. If an extremely high input resistance is necessary, another type of FET may be employed. This is the metal-oxide semiconductor FET or MOSFET, sometimes also referred to as a MOST or insulated-gate FET, or IGFET.

This device differs from the junction FET in that the gate is actually insulated from the channel by a very thin (about 100 nm) layer of oxide insulation, usually silicon dioxide (glass). The input resistance is then typically in the range 10^6 to 10^8 megohms. MOSFETs are described under two general types: the *depletion* type and the *enhancement* type.

Figure 3.13(a) and (b) illustrates the constructional features of the MOSFET. In both forms the gate and channel form the two plates of a capacitor separated by the thin silicon dioxide layer. Because the gate is insulated this way, V_{GS} can be either positive *or* negative with respect to the channel without conduction taking place through the gate-channel circuit. Any potential applied to the gate establishes a charge on the gate and this induces an equal but opposite charge in the channel.

In *Figure 3.13(a)*, when the gate potential is positive, a negative

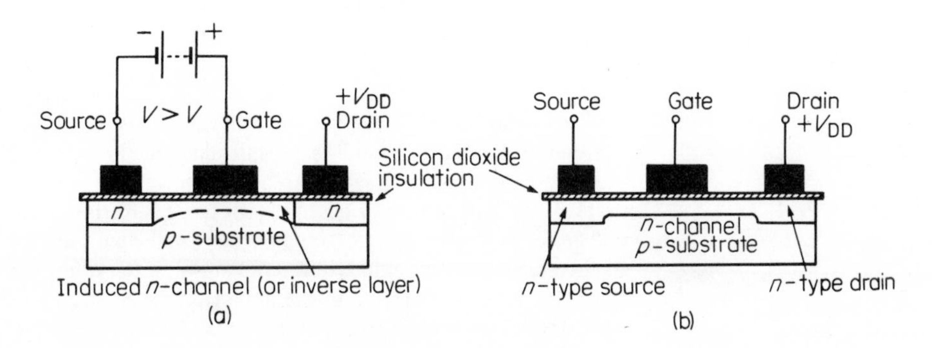

Figure 3.13

charge is induced in the *p*-type substrate at its interface with the silicon dioxide dielectric. This charge repels the majority carriers (holes) from the surface of the substrate and the minority carriers (electrons) that remain form an *n*-channel 'bridge' which connects together the existing *n*-type source and drain electrodes. Consequently, when the drain is connected to a positive supply voltage V_{DD}, electrons flow from source to drain by way of the induced n-channel. Increasing the gate-source voltage V_{GS} *positively* widens or *enhances* the induced channel and the flow of drain current increases. For this reason this form of MOSFET is known as an *n*-channel enhancement FET.

When drain current flows along the channel, a voltage drop is established and this tends to cancel the field set up across the dielectric by the positive gate bias. (Think about the workings of the JUGFET at this point). When the cancellation is sufficient almost to eliminate the induced *n*-channel layer, the channel pinches off and the drain current saturates at a value which is substantially independent of any increases in drain voltage. The output and mutual characteristic curves for an

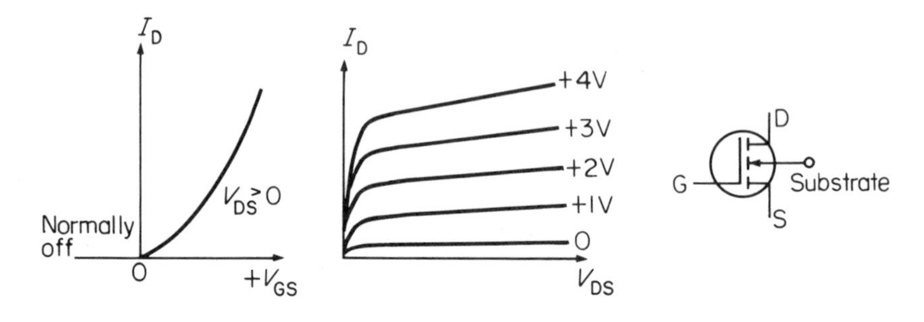

Figure 3.14

n-channel enhancement mode transistor are shown in *Figure 3.14* together with the circuit symbol. On the transfer characteristic the value of the gate-source voltage at which drain current falls to zero is called the threshold voltage V_T. It is that voltage which *just* induces the *n*-channel at the surface of the *p*-type substrate. The curve does not cross the I_D axis. The enhancement FET is a normally OFF device.

(9) Sketch on appropriately marked axes the characteristic curves of a *p*-channel enhancement type MOSFET.

Figure 3.13(b) shows the construction of an *n*-channel depletion mode MOSFET. Here the *n*-channel is introduced during manufacture by *n*-type doping of the surface layer of the substrate between the *n*-type source and drain regions. The construction is then essentially similar to the JUGFET except that the gate is insulated from the channel by the oxide layer. As a result, current flows from source to drain with zero gate voltage when a positive voltage is applied at the drain. If the gate is made *negative* with respect to the channel, the *n*-channel width is reduced (or depleted) as electrons are expelled from its interface with the oxide dielectric, and drain current decreases. The transistor is then operating in the depletion mode, just as the JUGFET does. Here, however, the gate voltage may be made positive, in which case the channel width is enhanced and drain current increases. Hence

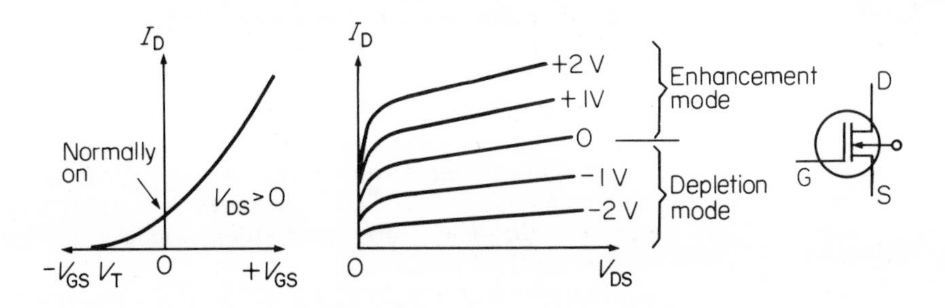

Figure 3.15

this type of MOSFET will operate in either the depletion or enhancement modes. The depletion FET is a normally ON device.

Figure 3.15 shows the output and mutual characteristics of a transistor of this kind. Notice that the threshold voltage now has a negative value (as for the JUGFET) and that the transfer curve crosses the I_D axis, where $V_{GS} = 0$ and $I_D = I_{DSS}$. The symbol for the depletion MOSFET is shown to the right of the diagram. The channel is now represented by a full line.

TEMPERATURE EFFECTS

Temperature variation leads to two effects in FETs: drain current I_D varies with temperature as shown in *Figure 3.16(a)*, and mutual conductance g_{fs} varies with temperature as shown in *Figure 3.16(b)*.

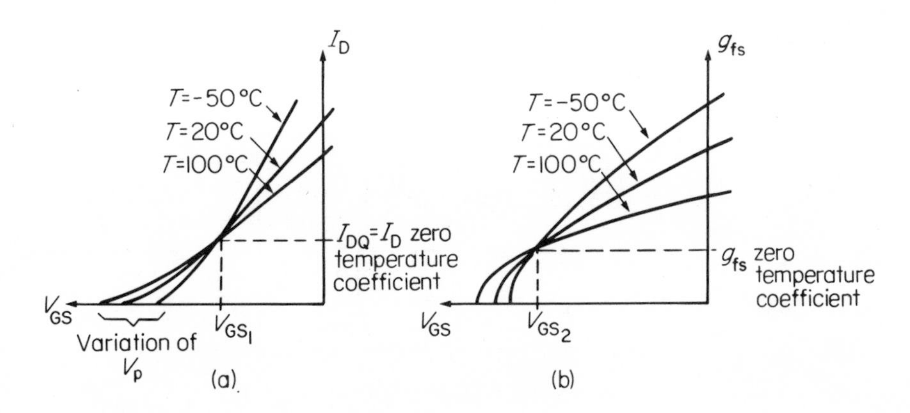

Figure 3.16

The variation in drain current comes about from two different effects. First, the resistivity of the channel increases with temperature, as do the majority of conductors, and for the same reason, thermal agitation reduces the velocity of the carriers, hence I_D decreases as temperature increases. Secondly, there is a variation in the gate-channel depletion layer with temperature change. As we know, the pinch-off voltage V_p is the gate potential required to theoretically just reduce the drain current to zero.

In fact, a very small leakage current continues to flow which varies with temperature. At room temperature this current is of the order of 50–100 pico-amps for *n*-channel transistors. The variation of the pinch-off voltage shown in *Figure 3.16(a)* is entirely due to the variation of the barrier potential (or contact potential) across the gate-channel *p-n*-junction, the same effect that causes the variation of V_{BE} with temperature in junction transistors. An increase in temperature will reduce the barrier potential and this will lead to a reduction in the width of the depletion layer for a given V_{GS}. The channel resistance will then fall and so I_D will increase.

You will have noticed that these two effects act in opposition, and it is possible to choose a particular V_{GS} to give zero temperature coefficient with respect to the drain current. Typically, a linear approximation of 2mV/°C for the variation of V_p with temperature is assumed, this value being negative for *p*-channel and positive for n-channel JUGFETS.

However as *Figure 3.16(b)* shows, there is also a variation in g_{fs} with temperature and there is also a bias point where g_{fs} has a zero temperature coefficient. This bias point, V_{GS2}, does not coincide with the one obtained for zero drain current temperature coefficient, V_{GS1}, but occurs when the bias is such that I_D is about $0.25 I_{DQ}$, see diagram (a). So a compromise is necessary in design. It can be shown that the best overall condition is obtained when $V_{GS} \simeq V_p + 0.63$ V, i.e. when the FET is biased 0.63 V above pinch-off.

EFFECTS OF INDUCTIVE LOADS ON FREQUENCY RESPONSE

Like the bipolar transistor amplifier, the FET amplifier has a gain-frequency response that depends upon the form of drain load, the inter-electrode and stray capacitances and the coupling system. When the drain load is a resistance, the response curve takes the general form shown in *Figure 3.17*.

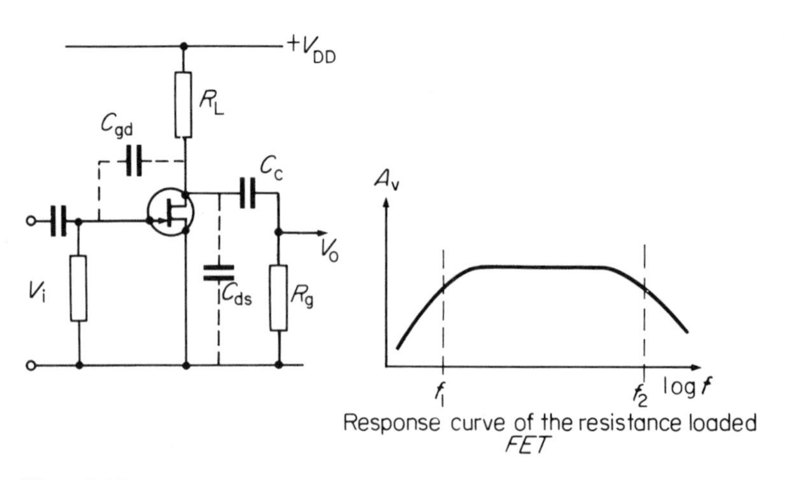

Figure 3.17

The fall in gain at high frequencies results from the shunting effect of the inter-electrode and stray capacitances upon the effective resistance of the load. The drain-source capacitance C_{ds} is effectively in parallel with R_L so reducing its load value at frequencies where the reactance of C_{ds} becomes comparable with or less than R_L; and the gate-drain capacitance C_{gd} effectively short-circuits the transistor itself at high frequencies. Typical upper frequency limits are of the order of 300 MHz for the JUGFET and some 100–15 MHz for MOSFET devices.

At the low frequency end of the spectrum the gain falls because of the increasing reactance of the coupling capacitor C_C relative to the input resistance of the following stage. This input resistance is shunted by R_g and when the reactance of C_C is comparable with or greater than R_g a large proportion of the signal is 'lost' across C_C.

The effect with the FET is considerably less than that experienced with the bipolar transistor, the reason being that the input resistance of the FET itself is extremely high and the total effective resistance even with R_g in shunt can still be of the order of a megohm or so. In the common-emitter stage the input resistance is only of the order of a few thousand ohms. Hence the value of C_C can, in the case of the FET, be relatively small for a given low frequency fall off in comparison with the large capacitance required in the bipolar case.

The upper frequency roll off point can be increased by the methods illustrated in *Figure 3.18(a) and (b)*, where an inductor L has been wired in the drain circuit. In diagram (a) the inductor has been added in series with the existing resistance load; the overall effect of this modification is that the increasing reactance of the inductor as the frequency increases compensates for the shunting effect of the capacitances upon the effective drain load.

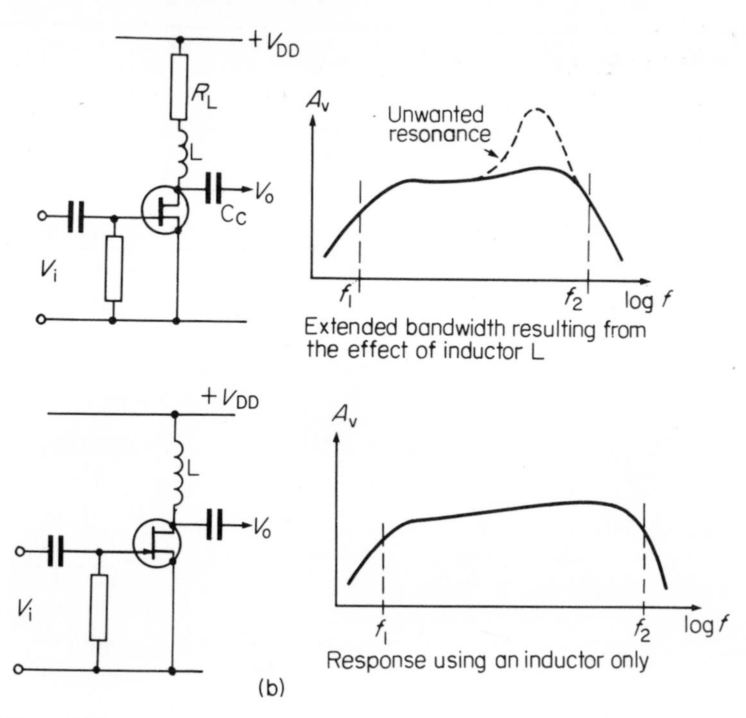

Figure 3.18

At some high frequency, L will resonate with the capacitances and the response curve will be boosted to a much higher frequency limit than would otherwise be obtained. The resonance must not be sharp, otherwise the response will tend to follow the broken line curve of diagram (a). This is undesirable, so the choice of inductor must be such not only that resonance occurs at the proper point but that it is sufficiently damped by the circuit resistances to give a flat response i.e. the Q of the inductor must be *low*.

In *Figure 3.18(b)* the load resistance R_L has been omitted and an inductor only is employed. The effect of this is that the gain is reduced at the low frequency end of the spectrum when the reactance of L is small, but increases with increasing reactance as the frequency increases. Again, resonance with the interelectrode capacitances can occur at some high frequency point but for a low-Q inductor this will extend the response without introducing an unwanted peak. You should compare this situation with transformer coupling mentioned in the previous chapter.

We assume that for all these response curves the input signal level is constant and that the voltage gain is given by V_o/V_i.

A TUNED DRAIN LOAD

If the inductor of *Figure 3.18(b)* is found to have a low self-resistance and hence a large value of Q, it will resonate with the circuit stray capacities to give a relatively sharp resonance curve at some high frequency. By introducing additional capacitance in parallel, the point at which resonance occurs can be brought under the control of the designer and a *selective amplifer* results, the response curve now being as shown in *Figure 3.19*, along with the circuit arrangement.

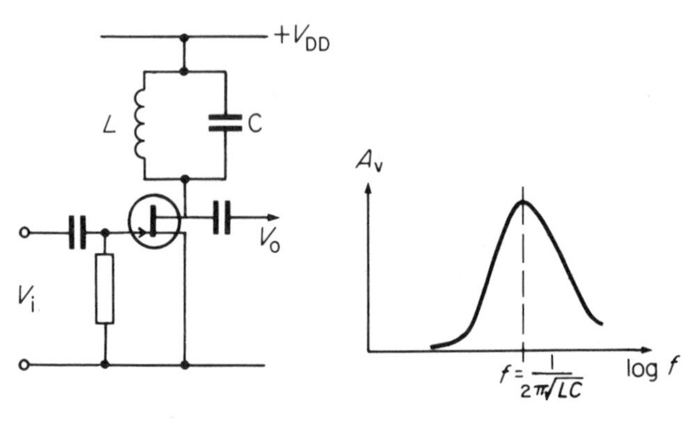

Figure 3.19

The voltage gain of this circuit is greatest at resonance and this depends upon the dynamic resistance R_D of the parallel tuned circuit. From equation (3.4) earlier, the gain becomes (for $R_D \ll r_d$)

$$A_v \simeq -g_{fs}.R_D \qquad (3.4)$$

where $R_D = L/CR$ ohms.

The frequency of resonance is given by $1/2\pi\sqrt{LC}$, but C here includes the inter-electrode and stray capacitances. Since we wish to tune the amplifier to some definite frequency, the presence of these strays will necessitate a slightly different value for the parallel capacitor than would be otherwise required. It is usual to make the added capacitance large in relation to the stray capacitance so that any slight variations in these latter are swamped out and the resonant frequency is not affected to any important extent.

THE FET AS A SWITCH

When analysing an active device such as a switch, the parameters of most interest are the switching times and the ON and OFF resistances. A perfect switch would have zero ON resistance and an infinite OFF resistance, and the time taken to switch from either state to the other would be zero. These conditions can never be achieved in the real world.

A mechanical switch comes very close to the ideal conditions for ON and OFF resistance, but is slow in its switching speed.

Another drawback is contact bounce (the contacting points oscillate for a very short period before becoming finally established) and there may be arcing, particularly when the contacts are broken. These failings would lead to serious problems in, say, computer circuits where positive switching times of fractions of a microsecond are necessary.

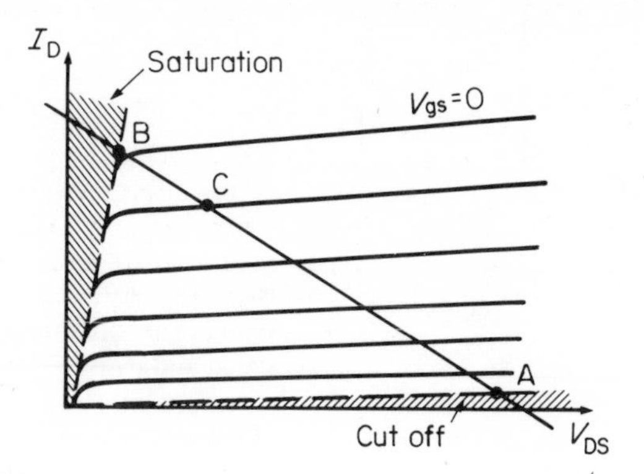

Figure 3.20

Figure 3.20 shows a load line drawn across the output characteristics of a FET. Point A represents the upper limit of the OFF region and point B represents saturation or full ON. Switching could occur between these extremes and should the gate voltage exceed zero volts to drive the drain current to point B or beyond, the transistor would be operating as a *saturating* switch. A *non-saturating* switch is one in which the gate voltage variations cause operation between the cut-off region at point A and some other point such as C. Point C represents an ON condition but not a saturated ON. A saturating switch requires a greater gate voltage drive but the ON resistance is much lower than it is in a non-saturating switch. All of what we have just said will, of course, apply equally well to a bipolar transistor, except that base current drive replaces gate voltage.

For high speed operation, any form of electronic switch is superior to a mechanical switch and since bipolar and FET devices can be used as such electronic switches it is of interest to compare their properties in this respect. A bipolar transistor has an ON resistance of about 150–200 Ω and an OFF resistance of some 100–200 Ω. It also has a very high switching speed, the limitation being due to the charge carrier storage within the two junctions which must leak rapidly for fast operation. Further, when the base current of a transistor changes abruptly, the collector current does not respond instantaneously.

Carriers from the emitter must travel across the base region and this takes a finite time known as the *transit time.* Hence there is a turn-on delay. In the FET the transit time is negligible compared to the bipolar transistor because of the much higher velocity of the charge carriers in the conducting channel.

In the FET the ON resistance ranges from 20 to 250 Ω and is given approximately by V_p/I_{DSS} in a saturating switch. This compares favourably with the bipolar transistor. The OFF resistance of the FET is much superior, however, being measured in hundreds of megohms.

The limiting factors of the FET switching speed are the interelectrode

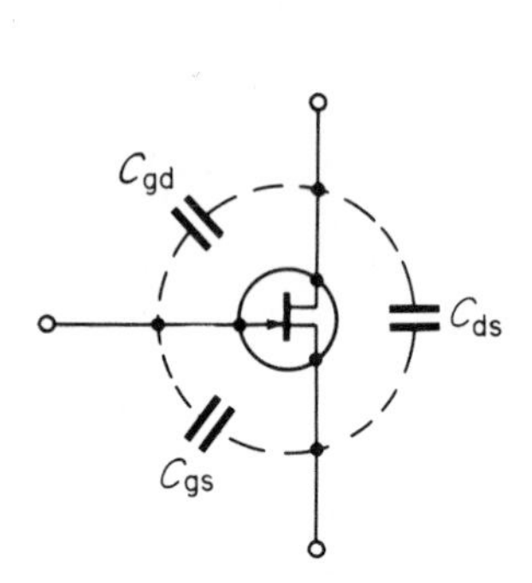

Figure 3.21

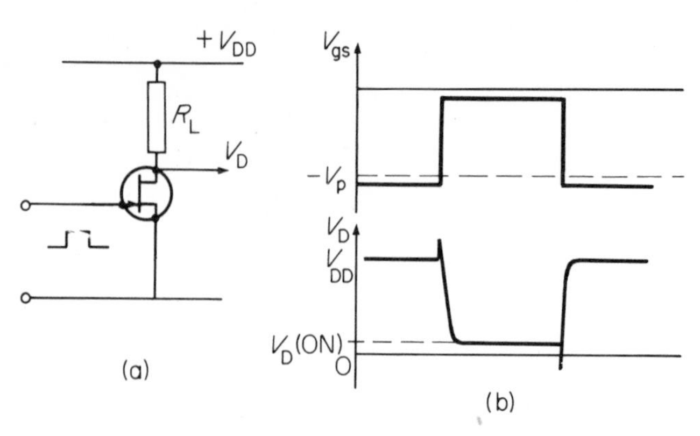

Figure 3.22

capacitances. *Figure 3.21* shows these capacitances. C_{gd} allows the gate signal to feed through to the drain, while C_{gs} reduces the switching speed and causes some 'loss' of signal as it is placed in parallel with the gate signal in the common-source connection. C_{ds} has a minimal effect but does contribute to some reduction in switching speed.

Figure 3.22(a) shows an n-channel FET connected as a switch. If a square wave input is applied at the gate, diagram (b), the transistor will switch ON when the signal carries the operating point to the $V_{DS(sat)}$ position and switch OFF when the signal reduces the input level to, or less than, the pinch-off figure. The operating point is then shifted to the bottom of the load line.

At saturation the voltage across the FET is not zero but is typically in the range 0.2 to 0.5 V. At switch-off, likewise, the drain current does not fall to zero but is typically about 1 nA for a JUGFET and 50–100 pA for a MOSFET. The output waveform will then be of the form illustrated, the initial spike resulting from the shunting effect of C_{gd} and the transition times, evident as sloping sides to the waveform, following from the effect of C_{gs}.

The output voltage in the ON condition, $V_{D(on)}$, can be calculated from

$$V_{D(on)} = \frac{\text{ON resistance}}{R_L + \text{ON resistance}} \times V_{dd}$$

The output voltage in the OFF condition is similarly given by

$$V_{D(off)} = \frac{\text{OFF resistance}}{R_L + \text{OFF resistance}} \times V_{DD}$$

Since $R_L \ll R_{off}$, this will simplify to

$$V_{D(off)} \simeq V_{DD}$$

Power dissipation in the FET depends upon the switching times between ON and OFF states. When the FET is switched on the voltage across it is very small and when it is switched off the current through it is very small. In either state the product (voltage × current) is small and so the power dissipation is small. Hence switching transitional periods determine the efficiency of the FET (as of any other electronic switch) as a switching element.

PROBLEMS FOR SECTION 3

(10) Sketch the construction and describe the principle of operation of a junction field-effect transistor. Explain what is meant by pinch-off and show its effect on the drain characteristics.

Draw a circuit diagram of an automatically biased junction FET used in a Class-A amplifier in the common-source mode. (C. & G.)

(11) Describe the principle and operation of either of the following insulated-gate field effect transistors: (1) depletion type, (ii) enhancement type. Sketch the symbols for these devices and draw typical gate/drain characteristic curves for both types.

(12) Define the parameters: (a) mutual conductance g_{fs}; (b) drain slope resistance r_d; (c) amplification factor μ. Given the relevant characteristic curves of a FET, how would you determine these parameters?

(13) A FET has a transconductance (mutual conductance) of 3 mS. If the gate-source voltage changes by 1 V, what will be the change in the drain current? What assumption have you made in obtaining an answer?

(14) When the drain voltage of a FET is reduced from 20 V to 10 V the drain current falls from 2.5 mA to 2.35 mA. Assuming the gate-source voltage remains unchanged, what is the drain slope resistance of the FET?

(15) The following values are taken from the linear portions of the static characteristics of a certain FET:

V_{DS}	12	12	8	volts
I_D	6.2	2.9	5.9	mA
V_{GS}	−1	−2	−1	volts

Estimate the parameters r_d, g_{fs} and μ.

(16) Complete the following statements:

(a) In general terms, FET's have properties most similar to . . . . .

(b) A JUGFET can operate only in the . . . . . mode.

(c) The V_{GS} polarity for an enhancement mode FET is dependent upon . . . . .

(d) I_{DSS} is the drain current which flows when V_{GS} is . . . .

(e) In a JUGFET majority carriers flow from . . . . . to . . . . .

(17) A FET has parameters g_{fs} = 2.0 mS, r_d = 50 kΩ. What is its amplification factor? If this FET is used as a common-source amplifier with a drain load of 10 kΩ, calculate the voltage gain.

(18) A FET used in common-source mode has g_{fs} = 2.0 mS, r_d = 100 kΩ. What value of load resistor will provide a voltage gain of 28 dB?

(19) In the circuit of *Figure 3.9*, V_p = −2 V and I_{DSS} = 2 mA. It is desired to bias the circuit to a drain current of 1.1 mA. If the drain supply is 15 V, estimate (a) V_{GS}; (b) g_{fs}; (c) source resistor R_S; (d) the required value of R_L to provide a voltage gain of 10.

(20) The output characteristics of a FET are given in the table. Plot these characteristics and determine the drain-source resistance from the characteristic for V_{GS} = −2 V.

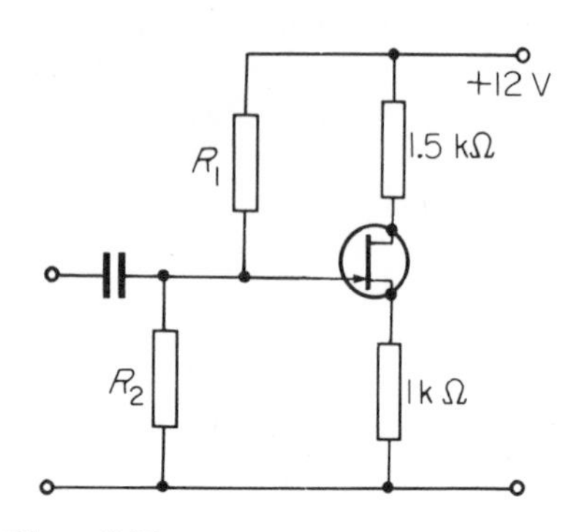

Figure 3.23

V_{DS} (V)	I_D (mA)			
	V_{GS} = 0	−1.0	−2.0	−3.0
2	5.6	4.15	3.0	1.90
6	7.0	5.25	3.62	1.91
9	8.1	6.0	4.1	1.92

Use the curves to estimate the g_{fs} of the FET for V_{DS} = 5 V.

(21) In the common-source amplifier of *Figure 3.23*, the quiescent conditions are such that $V_{GS} = -1.0$ V and the drain current I_D = 4 mA. Assuming the input resistance of the FET itself is infinite, obtain values of R_1 and R_2 such that the effective input resistance of the amplifier is 0.75 MΩ. What will be the approximate gain of the stage if g_{fs} = 5 mS? (*Hint:* R_1 and R_2 are effectively in parallel with the input terminals).

(22) The equation for the drain current I_D of a FET is

$$I_D = 7.5\left[1 - \frac{V_{GS}}{-2.5}\right]^2 \text{ mA.}$$

for a V_{DS} of 9 V. Find I_D for each of the following values of V_{GS}: 0, −0.5, −1, −1.5, −2 and −2.5 V. Plot the mutual characteristic for this FET and estimate g_{fs} when $V_{GS} = -1.5$ V.

(23) A FET is used as a common-source amplifier with a parallel tuned resonant circuit as the drain load. The tuned circuit consists of an inductor of 400 μH and resistance 20 Ω in parallel with a 300 pF capacitor. Calculate the resonant frequency.

If the transistor parameters are g_{fs} = 1.5 mS, r_d = 20 kΩ, calculate the voltage gain of the amplifier.

(24) Compare the properties of amplifiers using FETs with those using junction transistors.

A tuned amplifier has a FET with g_{fs} = 5 mA/V (mS) and r_d = 45 kΩ. The tuned circuit in series with the drain comprises a 140 μH coil in parallel with a capacitor. The amplifier has a maximum voltage gain of 185 at 800 kHz. Calculate (i) the capacitance, (ii) the Q-factor (iii) the voltage gain at 800 kHz when a 100 kΩ resistor is connected across the tuned circuit. (C. & G.)

(25) Discuss the factors that limit the switching speeds of (a) bipolar transistors, (b) field effect transistors.

A small transistor with a maximum power dissipation of one watt, can switch a load of several watts without damage. Explain why this is so.

4 Unwanted outputs: noise

Aims: At the end of this Unit section you should be able to:
Define noise as any unwanted signal.
List the sources of internal and external noise and explain how they occur.
Define signal-to-noise ratio and noise factor in an amplifier or receiver, and perform simple calculations relating to these definitions.
State the precautions taken to minimise the effects of external noise.
Explain a method of measuring noise factor.

Disconnect the aerial from your radio receiver (or tune it to a point between stations), turn up the volume control and put your ear to the loudspeaker. In theory you should hear nothing but you will in fact hear a background hiss while, if you listen carefully enough, will be heard to vary in intensity in a completely random fashion. You will hear the same sort of thing from any amplifier system which feeds into a loudspeaker, even though the input to the amplifier is completely disconnected.

This background noise is present even when the amplifier is working normally, though it tends to be concealed behind the speech or music output. Unlike distortion, this noisy background is present whether the signal itself is there or not. Other noises are also often present: there may be a background hum breaking through from the a.c. supply to the amplifier, crackling noises may be present particularly when thunderstorms are in the vicinity of the receiver or switches are operated in other rooms of the house. The list of such unwanted outputs is very long and all of them come under the heading of 'noise'. Noise need not even be aural, but visual. Noise appearing on a television signal, for example, shows itself on the screen as flickering spots or bands of varying light intensity.

We define 'noise' as any spurious or unwanted electrical signal set up in or introduced into an electronic system. Noise can be divided into two general classes:

(a) Internal noise generated within the electronic system as the result of the random movement of charge carriers in resistors, wires and active devices such as transistors and valves.

(b) External noise caused by atmospheric disturbances, diathermy apparatus, aircraft reflections and any spark-producing systems such as motor commutators or car ignition, to name just a few.

We will deal with these two classifications in turn.

INTERNAL NOISE

Although noise will be of concern at the output terminals of an amplifier or receiver, whether this output feeds to a loudspeaker, cathode-ray tube or any device on which the information may be recorded or displayed, it is in the early or small-signal stages that internally generated noise is of importance.

You will find that if you turn down the volume control on an amplifier system (in the absence of any signal), the background noise will become insignificant; so the large-signal stages which generally follow the position of the volume control are not contributing much of the noise you hear when the volume is increased. The bulk of the noise is appearing *before* the volume control, that is, in the small-signal stages. So the input stage of any amplifier or receiver is, in general, the vital area for investigation into noise generation, for the noise set up at this point determines the useful operational sensitivity of the unit as a whole. If the input signal magnitude is less than the noise level the output will be masked by the noise and become unintelligible.

Noise can be generated in any stage of an amplifier and from a variety of different processes. Many of these can be eliminated or minimised by careful attention to design and as such are not of primary interest to us in this section. For example, the proximity of an inductor, perhaps, carrying 50 Hz a.c. to the input terminals of an amplifier may introduce an unwanted hum signal into the amplifier. This is a case of stray inductive coupling which can be completely eliminated by magnetically screening the offending inductor and/or shifting its position and orientation relative to the sensitive input terminals. Stray capacitive couplings can likewise lead to the same trouble and can equally be eliminated by suitable screening and component disposition.

Other sources of internal noise are not, however, disposed of quite so easily. Internal noise is generated in resistors, transistors, valves and in fact every piece of wire used in the construction of an amplifier, and is a function of the physical properties of the materials used and the environment in which they operate. As a result such noise is substantially beyond the control of the designer to do anything about, at least, as far as layout is concerned.

Noise voltages of this sort are completely random in the sense that their future behaviour is unpredictable, in contrast to a sine wave, for example, whose past history is accurately known and whose future variations can be precisely stated. Noise voltages are not restricted to any particular frequency or phase or to any particular magnitude.

2. Thermal noise

Thermal noise, which is also known as Johnson noise, is the noise associated with the random movement of electrons in any electrical conductor. These movements, caused by thermal agitation, are present in every conductor even in the absence of an association with an electric circuit.

A piece of wire resting on a bench is generating thermal noise voltages. As the temperature of a conductor is increased from absolute zero, 0° K (= –273°C), the atoms of the conductor begin to vibrate

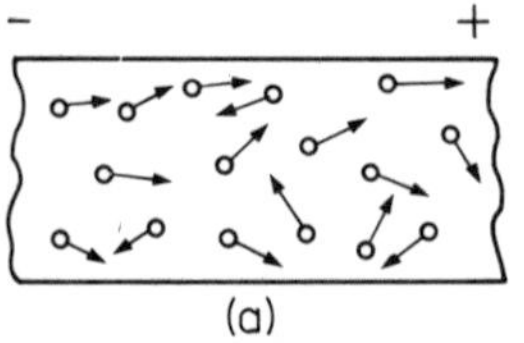

(a)
At this instant the direction of electron movement is predominantly left to right. The e.m.f. is set up with the polarity shown

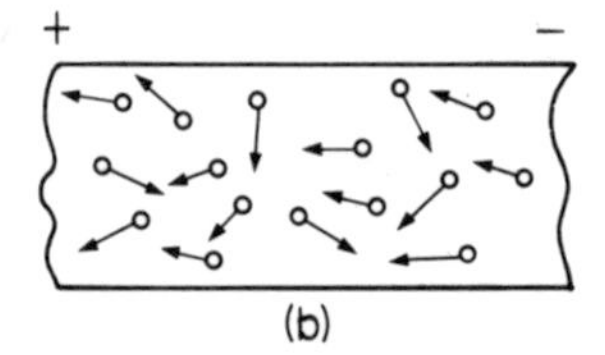

(b)
At this instant the direction of electron movement is predominantly right to left. The e.m.f. is set up with the polarity shown

Figure 4.1

and more and more electrons are gradually released from the outermost orbits of the atoms to become free charge carriers. These free electrons wander about within the atomic structure of the material first in one direction and then another, but at any given instant there will be more of them moving in one particular direction than another.

Since the movement of electrons constitutes an electric current, an e.m.f. will be instantaneously induced between the ends of the conductor in the manner shown in *Figures 4.1(a) and (b).* The time duration of each of these possible conditions is extremely short and so the polarity of the generated e.m.f. is fluctuating rapidly, as is its magnitude, and in a completely random way. This is thermal agitation noise.

Noise voltage is measured in terms of r.m.s. and is given by the relationship

$$V_n = \sqrt{(4kTRB)} \text{ volts} \tag{4.1}$$

where

k is a constant = 1.38×10^{-23} joules/°K;

T = the temperature of the conductor in degress K (= °C + 273°);

R = the resistance of the conductor in ohms;

B = bandwidth of the circuit in which the conductor is situated.

This last factor requires some additional explanation. Where noise is concerned, we must be interested in the *total* bandwidth of the system, so the whole of the area enclosed by the response curve must be considered. *Figure 4.2* shows a typical response curve where, in the usual way, the power gain is plotted against frequency. The equivalent noise bandwidth is obtained by constructing a rectangle ABCD which has an area equal to the area under the actual response curve. This can be done by estimating equal areas as shown in the diagram. The resulting noise bandwidth is *not* the same as the bandwidth obtained at the 3 dB points, but it is often sufficiently accurate to take the latter measurement as being the same as the noise bandwidth. Noise voltage depends upon the bandwidth of the circuit over which the noise is measured (or of the circuit at whose output the noise is present), whichever is the smaller.

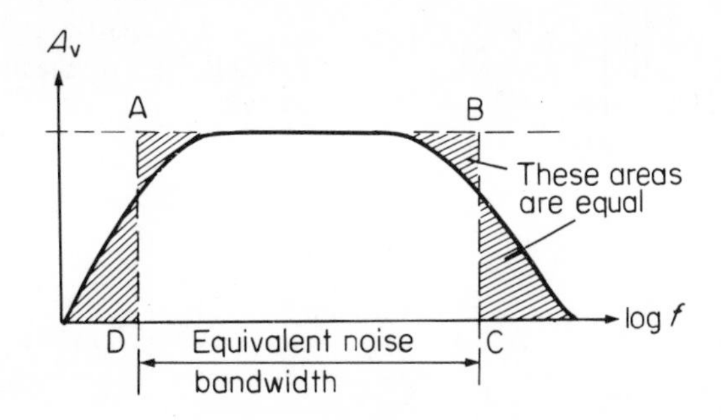

Figure 4.2

Thermal noise contains equal magnitudes of *all* frequencies and is known as 'white' noise. It clearly follows that we should expect a wideband amplifier to generate more noise than a narrow band amplifier.

Example (1) Calculate the thermal noise voltage generated by a 0.5 MΩ resistor at a temperature of 100°C if the bandwidth is 2 MHz.

$$V_n = \sqrt{4kTRB} \text{ volts}$$

where

$$k = 1.38 \times 10^{-23}; T = 100 + 273 = 373°K; R = 0.5 \times 10^6;$$
$$B = 2 \times 10^6.$$

$$V_n = \sqrt{(4 \times 1.38 \times 10^{-23} \times 373 \times 0.5 \times 10^6 \times 2 \times 10^6)}$$
$$= \sqrt{(2059 \times 10^{-11})} \text{ volts}$$
$$= 143\ \mu V.$$

A study of equation (4.1) tells us that to minimise thermal noise, the use of high value resistances operating at high temperatures must be avoided as far as possible. The temperature of a resistor is not simply that of the ambient but that resulting from its own generated heat. Strictly, equation (4.1) applies to metallic conductors and so to wire-wound resistors; carbon resistors introduce additional noise because the passage of direct current sets up very small arcs between the carbon granules making up the resistor body. Such internal *contact* noise has *no connection* with thermally generated noise. The use of low-noise resistors such as metal-oxide types, will minimize contact noise.

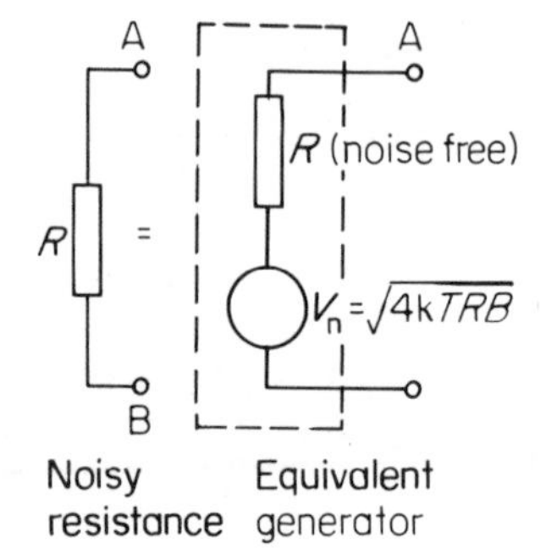

Figure. 4.3

A noisy resistor can be best represented for purposes of problem solving as a form of Thevenin equivalent generator as shown in *Figure 4.3.* Here the noisy resistance R is replaced by a noise-free ideal resistance of the same value, in series with a voltage generator developing an e.m.f. as given by equation (4.1). The open-circuit voltage across terminals A-B is then V_n. No polarity is associated with the hypothetical generator. We can now obtain an expression for the noise *power* developed in a resistance.

Let the equivalent noise generator of *Figure 4.3* be connected to a noiseless load resistor whose value is equal to the internal resistance of the source i.e. $R_L = R$, as shown in *Figure 4.4.* This condition, as you will recognise, represents the maximum power transfer from source to load. The voltage across the load will be one-half of the available voltage, that is

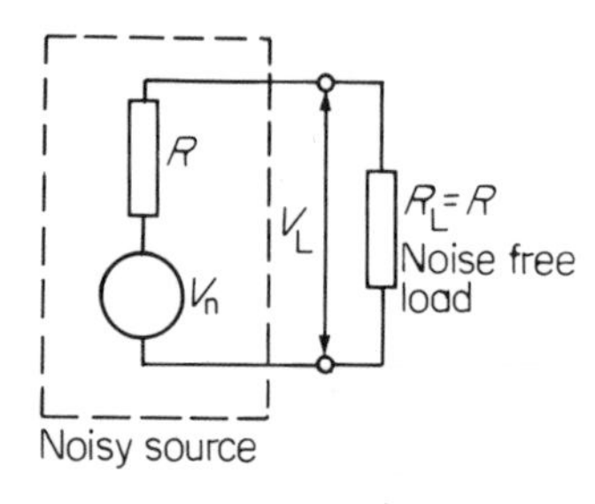

Figure. 4.4

$$V_L = \tfrac{1}{2}\sqrt{(4kTRB)} = \tfrac{1}{2}.2\sqrt{kTRB}$$
$$= \sqrt{kTRB} \text{ volts}$$

Hence the load power $P_L = \dfrac{(\text{load voltage})^2}{\text{load resistance}} = \dfrac{V^2}{R_L}$

since $\quad R_L = R.$

$$\therefore \quad P_L = \frac{kTRB}{R} = kTB \text{ watts} \tag{4.2}$$

Thus the *maximum* power that a thermal noise source can supply to a load when the source and load are matched is kTB watts. Notice that this power is *independent* of the value of the resistance but is *directly proportional* to *both* the temperature and the bandwidth.

When the source and load are not matched, this last equation is no longer valid and the noise power transferred to the load is reduced.

Some implications of what is written above may have occurred to you and caused puzzlement. For example, if we connect two resistors in parallel, which one supplies noise power to the other? Well, in the real world, the answer must of course be both. If the circuit containing the noisy resistor R is completed by loading it with another real (i.e. noisy) resistor of any value R_L, then at any given temperature the noise transfer of energy from R to R_L is exactly balanced by the noise energy transfer from R_L to R. Or, put in another way, over a period of time that is long compared with the average time that an electron travels in a particular direction, the total effective current flow in the circuit is zero. Which is what you would expect from just two resistors connected in parallel!

(2) Since the transfer of noise power is reduced when the load and source are not matched, would it be advantageous to build an amplifier with deliberately mismatched interstage couplings?

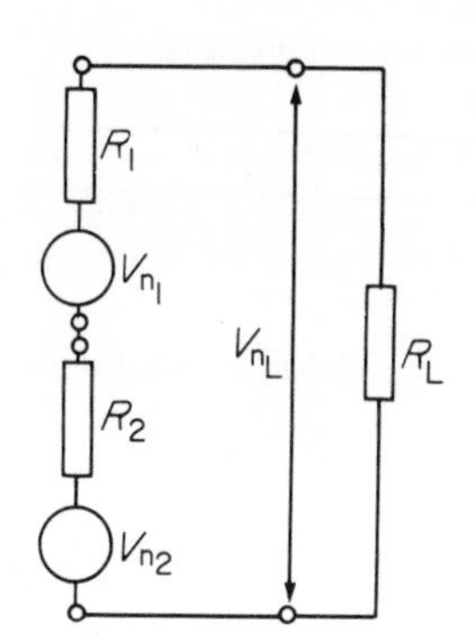

Figure 4.5

Generators in series. Suppose two noise generators are connected in series, what is the sum total of their equivalent noise output? One thing must be certain, there is no question of us simply adding voltages together *arithmetically*. Noise, remember, is completely random in frequency, phase and instantaneous magnitude. So both generators will set up currents through a load (see *Figure 4.5*) independently of each other and in a completely unrelated (uncorrelated) way. Thus the noise power in the load due to either generator is in no way influenced by the other. So the load noise power is

$$P_n = \frac{V_{n1}^2}{R_L} + \frac{V_{n2}^2}{R_L} = \frac{1}{R_L}\left[V_{n1}^2 + V_{n2}^2\right]$$

and the r.m.s. noise voltage across the load will be

$$V_{nL} = \sqrt{(R_L . P_n)} = \sqrt{(V_{n1}^2 + V_{n2}^2)}$$

Hence noise voltages add as the *square root* of the sum of the squares. In other words, V_{nL} can never be zero as would be possible if we just added the voltages arithmetically.

The following internal noises are all associated with transistors and valves.

3. Shot noise

Shot noise in an active device is the result of fluctuations when charge carriers cross potential barriers. In the thermionic valve the noise results from fluctuations in cathode current due to random variations in the cathode space charge. In the transistor, shot noise is caused by the random arrival and departure of carriers by diffusion across a *p-n* junction. A transistor has two such junctions and hence two such sources of noise. Shot noise is most important when low-level signals are being amplified, and in general it is more significant than the thermal noise of associated components.

3. Partition noise In a tetrode or pentode valve, the screen grid takes current as well as the anode. There is therefore a division of cathode current between these two electrodes. The screen current has random fluctuations since it is derived from the cathode, and these fluctuations are also superimposed (in the opposite phase) on the anode current.

In the transistor, the input current flows through the emitter to the base-emitter junction, after which it divides between the base and collector circuits, since $I_E = I_B + I_C$. The base current has random fluctuations and these are superimposed on to the collector current. Partition noise is generally negligible in modern transistors, and is not present in triode valves.

4. Flicker noise The cause of flicker noise is not fully understood but it is believed to be due to irregularities in the cathode coating in valves and to emitter surface leakage and recombination in transistors. This noise is predominantly low frequency in nature and becomes insignificant above a few kilohertz. Unlike white noise, flicker (or 'pink) noise departs from the flat relationship with frequency and becomes frequency dependent. For this reason it is often known as $1/f$ noise. Sensitive circuits for low frequency amplification are particularly affected by the presence of $1/f$ noise which appears in the FET as well as the bipolar transistor.

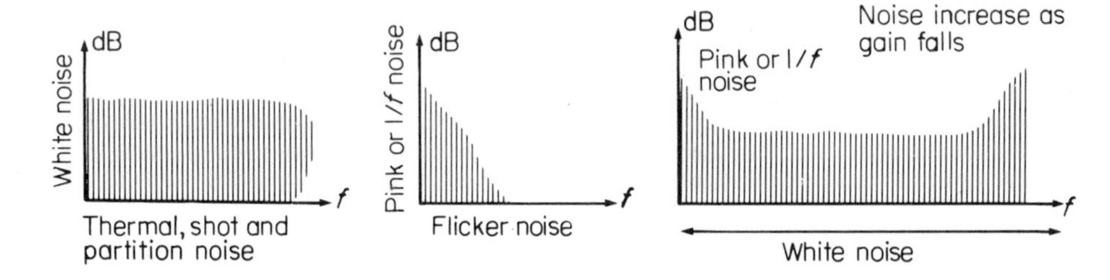

Figure 4.6

Figure 4.6 compares the spectral densities of white noise and pink noise and shows their overall resultant. You should note that noise increases above the white noise level as higher frequencies are approached because of the fall in gain of the system.

Example (3) Why is a field-effect transistor likely to be less noisy, under the same operating conditions, than a bipolar transistor?

Like the bipolar transistor, the FET exhibits shot noise, thermal noise and $1/f$ noise. However, the FET is inherently a lower noise device than an ordinary transistor because:

(a) There is only one *p-n* junction involved, so shot noise is reduced;

(b) There is no equivalent base (gate) current and so the source current does not divide before reaching the drain. Partition noise is consequently absent;

(c) Channel resistance is small, hence thermal noise is small.

(4) If a transistor is operated with a very small collector current, will the shot noise be greater, less or the same as it is when the collector current is large? Give reasons for your answer.

EXTERNAL NOISE

Much of the noise originating outside of any electronic system is man-made and is, therefore, under control to a certain extent. Other sources contribute natural noise and these are not under control.

MAN-MADE NOISES

1. Mains hum

Any electronic system that receives its supply voltage(s) from a rectifier type mains power unit or uses thermionic valves whose heaters are run from low voltage a.c. supplies, is liable to have hum introduced into its circuitry. Alternating magnetic flux from wires carrying alternating current or from the iron cores of transformers and chokes used in the power units, may link with the input circuits of the amplifying devices and induce there an unwanted e.m.f. Similarly, stray capacitance between active terminals and some other conductor around which there is an alternating electric field will lead to the flow of unwanted current in the active leads.

Electric and magnetic screening, together with a sensible layout and orientation of components will, as mentioned earlier, overcome most of the problems associated with this kind of noise. Hum signals introduced on the actual d.c. supply leads can be eliminated by additional smoothing and decoupling or by better regulation (stabilisation) at the power unit.

2. Spark or transient interference

Whenever an electric circuit is interrupted, whether by a switch, motor commutator, car ignition or welding equipment to name a few, the resulting sparks generate a high-frequency current in the circuit wires and these in turn act as transmitting aerials, radiating radio signals which cover a very wide band of frequency. There are also voltage pulses set up which pass along the circuit wires in which the switching takes place.

Ideally, when a circuit is interrupted, the current should fall instantly to zero (*Figure 4.7(a)*) but because of the presence of inductance and capacitance it is possible for a resonant condition to exist which can be shocked into oscillation as soon as the current changes, *Figure 4.7(b)*. The resulting interference can get into an electronic system either by way of the radio-frequency signals being picked up directly on the system wiring or aerial array or by way of the mains input connection.

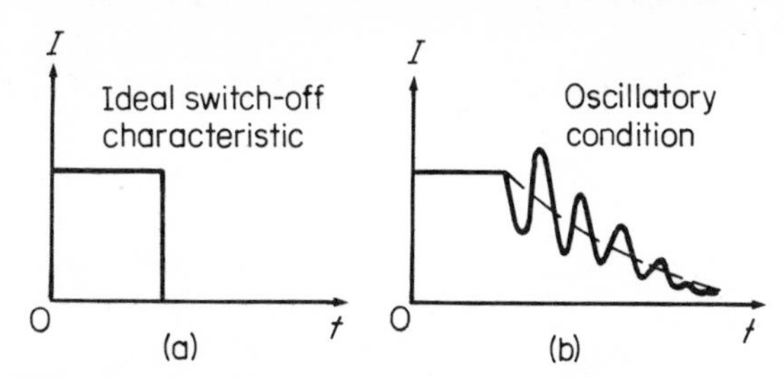

Figure 4.7

Screening the equipment will prevent direct radiation pick up and choke-capacitance suppressors wired into the mains lead will eliminate most of the voltage pulses coming that way. *Figure 4.8* shows a typical mains suppressor circuit. Aerials cannot be screened of course, but they can be orientated away from the source of the interference (if this is

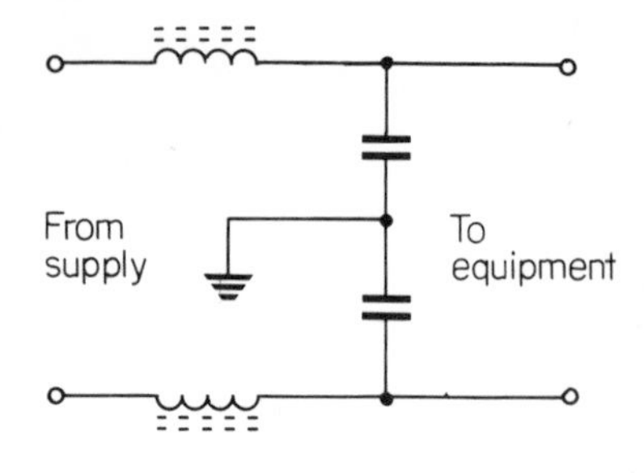

Figure 4.8

coming from a fixed direction) or re-positioned so that some sort of localised screening is achieved, such as the wall of a house of a line of trees being interposed between the aerial and the source of interference. Raising the aerial and screening the downlead often helps. The apparatus responsible for the interference is the logical place for screening and suppression systems to be fitted.

Ignition system interference is particularly troublesome to car radio receivers and nearby television receivers. Car radio receivers must always be properly earthed to the car chassis and both the aerial and the supply voltage leads must be screened.

3. Natural noise The bulk of natural noise is contributed by atmospheric or *static* noise resulting from lightning discharges. The effect of lightning on radio reception is well known, resulting either in violent crackling sounds or, in the presence of 'sheet' lightning, an almost continuous 'frying' background to the wanted programme. Atmospheric noise of this sort can be picked up over very great distances and static interference is often detectable even when no thundery weather is apparent.

At higher frequencies, above some 20 mHz or so, the effect diminishes but other sources of noise begin to show themselves. Solar disturbances lead to background noise and fading at frequencies above some 50 MHz, and thermal radiation from the surface of the earth, particularly in the summer months, generates noise which becomes prominent at frequencies over 200 MHz. Variations in the ionized layers in the upper atmosphere lead to what is known as quantum noise, though only very high frequency bands are affected by this. Noise sources in the Milky Way (which are put to good effect in radio astronomy) can be a nuisance to others on frequencies above some 1000 MHz.

SIGNAL-TO-NOISE RATIO AND NOISE FACTOR

We have already noted that the noise generated in an amplifier or introduced along with the input signal determines the smallest signal which can usefully be amplified. So important measures of the performance of an amplifier or receiver are what we call the *noise factor* (sometimes the noise figure) and the *signal-to-noise* (S/N) *ratio*.

Signal-to-noise ratio is defined as the ratio of the wanted signal power to the unwanted noise power, so

$$\text{S/N ratio} = \frac{\text{signal power}}{\text{noise power}} \qquad (4.3)$$

As it is usual to express this ratio in decibels, we have

$$\text{S/N ratio} = 10 \log \frac{\text{signal power}}{\text{noise power}} \text{ dB} \qquad (4.4)$$

Since $P = V^2/R$, if we substitute into equation (4.3) for the case where the signal voltage is developed in series with a total resistance R, we get

$$\frac{\text{signal power}}{\text{noise power}} = \frac{(\text{signal voltage})^2}{\text{R}} \times \frac{\text{R}}{(\text{noise voltage})^2}$$

$$= \left[\frac{\text{signal voltage}}{\text{noise voltage}}\right]^2$$

$$\therefore \quad \text{S/N ratio} = 10 \log \left[\frac{\text{signal voltage}}{\text{noise voltage}}\right]^2$$

$$= 20 \log \left[\frac{\text{signal voltage}}{\text{noise voltage}}\right] \qquad (4.5)$$

Audio amplifiers usually quote signal-to-noise ratio at a frequency of 1 kHz which then states an average value, but separate figures are sometimes given for both the low and high frequency ends of the operational range. Typical figures would be 60 dB for amplifiers and 55 dB for stereo f.m. radio receivers. In some systems, figures down to 15 dB are acceptable.

Example (5) If the signal level at the input of a radio receiver is 20 μV and the noise level is 1.5 μV, calculate the S/N ratio in dB.

$$\text{S/N ratio} = 20 \log \left[\frac{\text{signal voltage}}{\text{noise voltage}}\right]$$

$$= 20 \log \left[\frac{20}{1.5}\right] = 20 \log 13.34$$

$$= 22.5 \text{ dB.}$$

Now try the following problems on your own.

(6) If the signal level at a tape recorder head output is 1 mW and the signal-to-noise ratio is 40 dB, what is the output noise power?

(7) If the amplifier following the above tape head has a gain of 35 dB, what will be the S/N ratio at the amplifier output, assuming *no more noise* is introduced?

(8) An amplifier has a S/N ratio of power equal to −10 dB at its output when its bandwidth is 10 kHz. What must be the bandwidth so that a +5 dB S/N power ratio is obtained? [*Hint: the noise is reduced 15 dB*].

The measure of the noise quality of an amplifier is the *noise factor*. The noise factor F of an amplifier is the ratio of the total noise at the output to that part of the output noise which is due to the input noise only, that is

$$\text{Noise factor } F = \frac{\text{total noise power at the output}}{\text{output noise power due to input noise power}}$$

It is therefore a measure of the amount of internal noise introduced by the system per unit gain. If the amplifier introduced no noise, F would be unity. Naturally, we should aim to keep F as small as possible.

Figure 4.9 shows an amplifier having a power gain A_p and an internally generated noise power P_{na}. This amplifier is fed from a matched

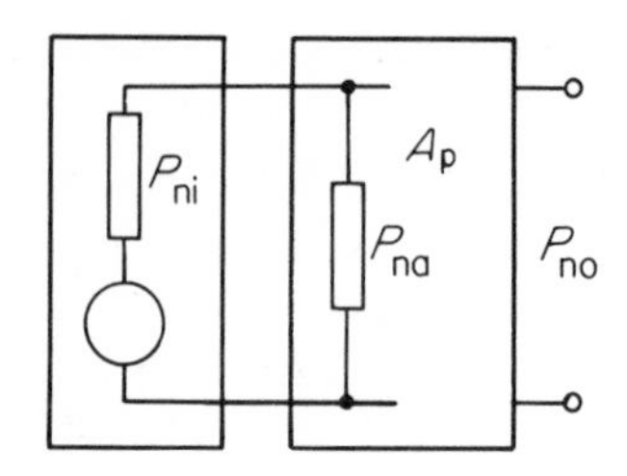

Figure 4.9

source which provides an input noise power P_{ni}. The total noise power at the output is P_{no}. Then

$$F = \frac{P_{no}}{P_{ni} \times A_p} \tag{4.6}$$

But $$A_p = \frac{\text{signal power at the output}}{\text{signal power at the input}} = \frac{P_{so}}{P_{si}}$$

$$\therefore \quad F = \frac{P_{no}}{P_{ni} \times P_{so}/P_{si}} = \frac{P_{si}/P_{ni}}{P_{so}/P_{no}}$$

$$= \frac{\text{input signal-to-noise ratio}}{\text{output signal-to-noise ratio}} \tag{4.7}$$

Now equation (4.6) can clearly be written in the form

$$F = \frac{P_{no}}{kTB \times A_p}$$

since the maximum power that a thermal noise source can supply to a matched load is kTB watts (from equation (4.2)). Then $P_{no} = FkTBA_p$. This output noise power is made up of amplified input noise ($kTBA_p$) and amplified internal noise, so that the internal noise power, P_{na}, is the difference given by $P_{no} - kTBA_p$, so

$$P_{na} = FkTBA_p - kTBA_p$$

$$= (F - 1)kTBA_p \tag{4.8}$$

Let the amplifier shown in *Figure 4.10* be made up of an input stage having a noise factor F_1 and power gain A_{p1} and the subsequent stage or stages having a noise factor F_2 and power gain A_{p2}. We assume that the useful bandwidth B is limited by the subsequent stages. The input provides a noise power kTB watts and after amplification this appears at the output as $kTB \times A_{p1} \times A_{p2}$ watts.

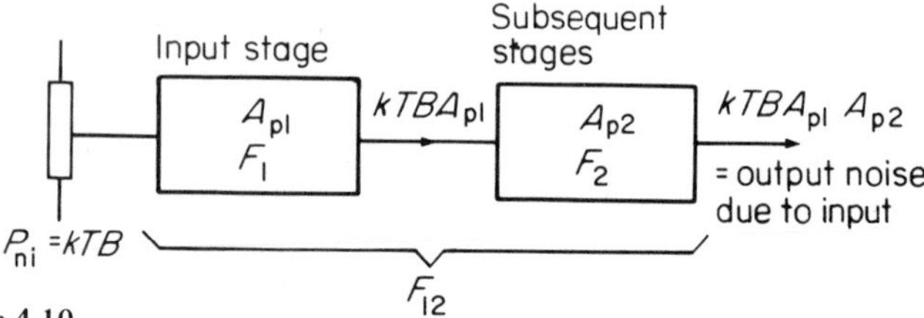

Figure 4.10

From equation (4.8) the internal noise from the input stage is $(F_1 - 1)kTBA_{p1}$ and this appears at the output as $(F_1 - 1)kTBA_{p1} \times A_{p2}$. Now the *overall* noise power originating from the input is

$$F_{12} = \frac{\text{total noise output}}{\text{noise at output due to input}} = \frac{\text{total noise output}}{kTBA_{p1}A_{p2}}$$

$$= \frac{kTBA_{p1}A_{p2} + (F_1 - 1)kTBA_{p1}A_{p2} + (F_2 - 1)kTBA_{p1}A_{p2}}{kTBA_{p1}A_{p2}}$$

$$\therefore \quad F_{12} = F_1 + \frac{F_2 - 1}{A_{p1}} \tag{4.9}$$

A study of this last equation reveals that the overall noise factor of a cascaded amplifier is primarily influenced by the noise introduced in the first stage. To make F_{12} low, F_1 must be low and the first stage gain A_{p1} must be very large.

Example (9) An aerial has a noise power equal to 0.1 μW and the noise factor of a receiver matched to this aerial is 10. Calculate (a) the total noise power at the output of the receiver if it has a power gain of 80 times, (b) the proportion of the output noise power contributed by the receiver.

(a) From equation (4.6) output noise power $P_{no} = F \times P_{ni} \times A_p$

$= 10 \times 0.1 \times 80\ \mu W$

$= 80\ \mu W$

(b) The noise power at the output due to the input noise

$= 0.1 \times 80 = 8\ \mu W$

∴ the amplifier contributes $80 - 8 = 72\ \mu W$ of noise

$= 90\%$ of total noise.

(10) A receiver has a power gain A_p, a bandwidth B Hz and a noise factor F. Prove that the noise power introduced by the receiver can be expressed as $(F - 1)kTBA_p$ watts.

Bear in mind that although we have so far expressed noise factor F as a numerical ratio, it can be expressed in dB in the usual way. A typical figure for a communications receiver would be 6 dB. A figure of 3 to 4 dB would be considered very good.

PROBLEMS FOR SECTION 4

(11) Give an example of (a) a low-frequency noise, (b) a high-frequency noise, (c) a man-made noise, (d) a natural noise.

(12) What is the noise voltage of a 100 k resistor at 27°C if the bandwidth is 1 MHz?

(13) If the signal level at the input to a receiver is 20 μV and the noise level is 1 μV, what is the signal-to-noise ratio in dB?

(14) The signal level from a microphone output is 5 mW. If the S/N ratio is 40 dB, what is the output noise power?

(15) Why is a FET less noisy than a bipolar transistor?

(16) What is meant by the terms 'white' noise, 'pink' noise and $1/f$ noise?

An amplifier has a flat passband characteristic extending from 100 kHz to 500 kHz and a gain of 40 dB. What will be the noise voltage at its output resulting from a 500 kΩ resistor at 20°C connected across its input terminals?

(17) If the effective noise level at the input to a receiver is −130 dB relative to 1 mW and the signal-to-noise ratio expected is 23 dB, what receiver gain is required to produce an output of 25 mW? (*Note*: a dB level referred to 1 mW is symbolised as dBm). (C. & G.)

(18) A receiver is receiving a signal and measurements show that the S/N ratio at the receiver output is 30 dB. Assuming that all the noise occurs in the first stage of the receiver, calculate the

output S/N ratio when (a) the output stage gain of the receiver is increased by 6 dB; (b) the incoming signal fades so that the receiver power falls to one quarter of its previous value.

(19) Prove that if an amplifier has a gain G and a noise factor F, then the product $G(F-1)$ is proportional to the internal noise.

(20) What is meant by the noise factor of an amplifier?

Prove that the power delivered to a matched load by a noise source of internal resistance R is kTB watts.

A communications receiver has a noise factor of 10 dB at 30 MHz. Is this a poor, typical or very good figure?

(21) If the noise power in a given bandwidth in an aerial is 25 pW (1 $p\mathrm{W} = 10^{-12}\,\mathrm{W}$) and the noise factor of a receiver matched to the aerial is 10, calculate (a) the total noise power in the given bandwidth at the output of the receiver if it has a power gain of 50; (b) the noise power at the output due to the input noise; (c) the contribution of the amplifier to the output noise power.

What would happen, if anything, to the noise power if the bandwidth was halved?

(22) An amplifier has a gain of 20 dB and a 6 dB noise factor. An aerial produces an available thermal noise power in a given bandwidth of -110 dBm. If the aerial is attached to the amplifier input and both are at the same temperature, in a given bandwidth what is (a) the total noise power at the output; (b) the noise power at the output due to the amplifier noise only?

If an identical amplifier is cascaded with the first, what will be the noise factor of the combination?

5 Negative feedback

Aims: At the end of this Unit section you should be able to:
Understand the basic principles of negative feedback.
Differentiate between series- and parallel-derived feedback methods.
Explain the effect of negative feedback on amplifier gain, bandwidth, noise, distortion and input and output resistance.
Apply feedback principles to practical amplifier circuits.

Many of the characteristics of amplifiers which we have so far treated as fixed and unvarying, such as gain and gain stability, bandwidth, input and output resistances, are actually at the mercy of variations in circuit parameters and environmental conditions. Transistor gain tolerances, operating point shift, the ageing of components, temperature and climatic changes, supply voltage variations, to name only a few, all tend to, and do, affect an amplifier adversely.

The effects of most of these variations can be eliminated or at least considerably reduced by the introduction of controlled *feedback* which, as the name suggests, involves the addition of a portion of the amplifier output signal to the input signal. When the input signal is effectively reduced in magnitude by the addition of the feedback, the method is known as *degenerative* or *negative* feedback. If there is an increase in the total input due to the feedback, the method is termed *regenerative* or *positive* feedback. If an amplifier is badly designed unintentional feedback can occur because of coupling between electric or magnetic fields in different parts of the circuit. Such feedback is invariably of the regenerative kind and the amplifier is quite unsuitable for its intended function in life. Controlled feedback can only be usefully applied to an amplifier *which has been soundly designed in the first place.*

The use of negative feedback is widespread but it must be used with caution. Negative feedback significantly modifies the characteristics of an amplifier to which it is applied and enables a predictable performance to be achieved, free from the effects of the variations in the circuit components and devices from which it is constructed. It also ensures repeatibility, every amplifier in a manufactured batch having identical characteristics and lying within tight specification limits.

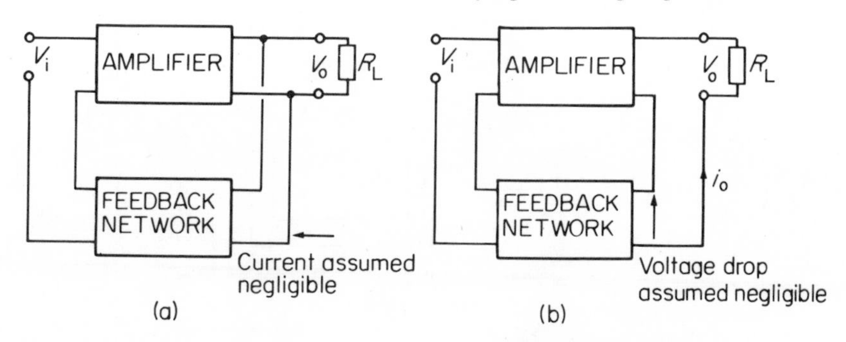

Figure 5.1

We shall be concerned in this Unit section with amplifier systems in which the output signal is fed back to the input terminals in antiphase with the input signal. This is negative feedback (n.f.b.). Cases of positive feedback in which the signal is fed back in phase with the input will be covered in the section 6.

The manner in which the output signal is sampled and the manner in which it is introduced into the input circuit leads to four basic classifications of n.f.b. amplifiers. Two of these are shown in *Figure 5.1(a) and (b)*. These illustrate those cases where the voltage fed back is applied in *series* with the input. In diagram (*a*) the voltage fed back from the output is proportional to the output voltage v_o and is derived from a network in parallel with the load resistor R_L; this is known as *voltage-voltage* feedback. In diagram (*b*) the voltage fed back and applied in series with the input is proportional to the current flowing in the load resistor, i_o. This is known as *voltage-current* feedback, although it is a voltage and not a current which is actually fed back.

We consider first the general principles involved when the feedback is applied in series with the input signal, irrespective of the form of connection at the output end.

VOLTAGE FEEDBACK PRINCIPLES

In voltage feedback, the signal fraction fed back is proportional to the voltage of the signal at the output of the amplifier, or at some point within the amplifier at which the feedback signal is derived. *Figure 5.2(a)* shows an amplifier with voltage gain A_v supplying a load resistor R_L. As we know

$$\text{Voltage gain } A_v = \frac{v_o}{v_i}$$

or

$$v_o = A_v . v_i$$

The diagram of *Figure 5.2(b)* illustrates the same amplifier with voltage feedback. The basic amplifier gain is still A_v and the output voltage v_o is developed as before across R_L. In parallel with R_L is a feedback network consisting of two series-connected resistors R_1 and R_2. We assume that the total resistance of R_1 plus R_2 is so large compared with R_L that the current through them is negligible. That part of the output voltage developed across R_2 is fed back and added in series with the input voltage; let this fraction be called β, so that

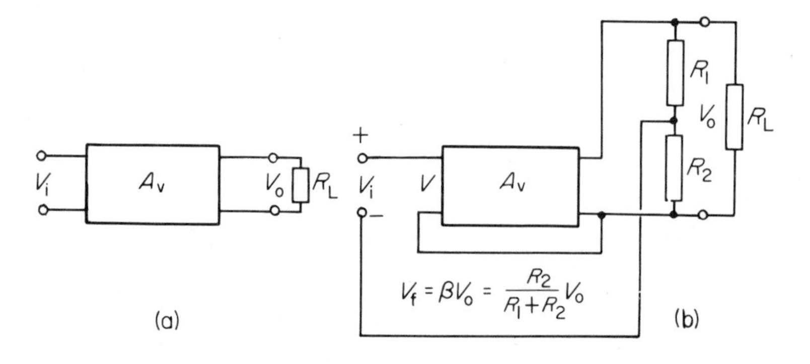

Figure 5.2

$$\beta = \frac{R_2}{R_1 + R_2}$$

Then the voltage fed back, $V_f = \beta v_o$ and the total input voltage now available at the *terminals* of the amplifier is

$$v = v_i + \beta v_o$$

But now

$$A_v = \frac{v_o}{v} \text{ or } v = \frac{v_o}{A_v}$$

$$\therefore \quad \frac{v_o}{A_v} = v_i + \beta v_o$$

and so

$$v_o = A_v . v_i + A_v \beta v_o$$

Hence the gain with the feedback applied, measured between the *new* input terminals and the output is

$$\frac{v_o}{v_i} = A_v{}' = \frac{A_v}{1 - \beta A_v} \tag{5.1}$$

This is the *general feedback equation.*

The quantity βA_v in this equation is called the *loop gain* since it is the total gain measured round the feedback loop from v to V_f i.e. it is the product of the individual gains of the amplifier, A_v, and the feedback network β. In the absence of feedback, $\beta = 0$ and the amplifier exhibits a gain equal to A_v. This is the *open-loop* gain.

There are three possible forms that the loop gain product may take which are of interest:

1. If βA_v lies between 0 and 1 (but not equal to 1), the denominator of equation (5.1) will be *less* than unity and the gain with feedback, $A_v{}'$ will be *greater* than A_v. This is a case of positive feedback because the gain of the system is increased.
2. If βA_v is negative, the denominator will be *greater* than unity since $1 - (-\beta A_v) = 1 + \beta A_v$, and $A_v{}'$ will be less than A_v. This is a case of negative feedback because the gain of the system has been reduced.
3. If $A_v = 1$ the denominator becomes zero and the resultant gain is theoretically infinite. This implies that the circuit has an output independent of an external input voltage; hence the circuit will be unstable and oscillate. We shall consider this special condition in the appropriate section. Strictly, equation (5.1) does not apply in this case, since it was derived on the assumption that the amplifier system of *Figure 5.2(b)* was stable.

The general feedback equation is very simple to use and is easily remembered.

Example (1) In a certain amplifier the open-loop gain $A_v = 200$. Find the overall gain with feedback when (a) $\beta = 0.004$, (b) $\beta = -0.02$.

$$\text{(a) } A_v{}' = \frac{A_v}{1 - \beta A_v} = \frac{200}{1 - (0.004 \times 200)}$$

$$= \frac{200}{0.2} = 1000$$

$$\text{(b) } A_v' = \frac{200}{1-(-0.02 \times 200)} = \frac{200}{5} = 40$$

Clearly, the introduction of negative feedback as part (b) of this example illustrates has led to a drastic reduction in gain. This will always be the case if βA_v is negative. This might seem at first to be a considerable handicap, but the advantages of feedback are so numerous that it pays to design the original amplifier with sufficient gain so that after feedback has been added, the overall gain will not fall below the desired figure.

PHASE SHIFT REQUIREMENTS

Let us take a more general view of an amplifier and a feedback network; this is illustrated in *Figure 5.3*. For βA_v to be negative, either A_v or β must be negative. Now either the amplifier or the feedback network, or both, may introduce a phase shift between their input and output terminals. These phase shifts have been respectively indicated by the angles θ and ϕ in *Figure 5.3.*

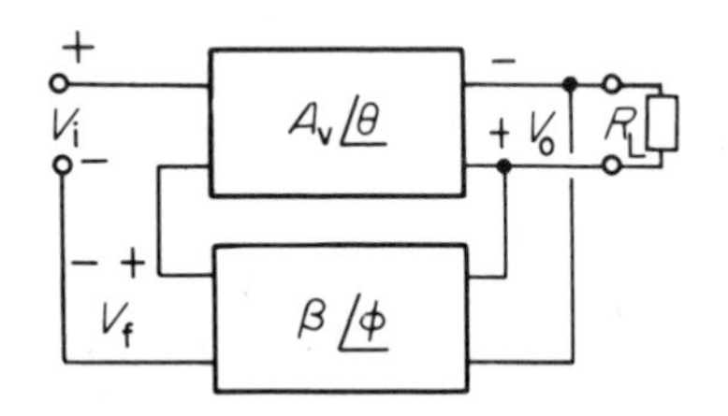

Figure 5.3

In our previous diagram of *Figure 5.2(b)* the feedback network consisted simply of a resistance divider, hence there was no phase shift there and angle ϕ was zero. For βA_v to be negative, therefore, the amplifier gain must be $-A_v$, hence the amplifier phase shift θ must be 180°. The total phase shift around the loop is then 180° and so the voltage fed back to the input (V_f) is antiphase to the existing input. The signal polarity signs marked on *Figure 5.3* show how this occurs.

If the amplifier consisted, say, of two common-emitter stages, its total phase shift would be 360° which is equivalent to a zero phase shift, hence $\theta = 0°$. In order to provide negative feedback for this amplifier, the phase shift in the feedback network, ϕ, would have to be 180°.

To simplify matters we will assume throughout this section that βA_v is negative and not worry ourselves whether it is the amplifier or the feedback network which is actually contributing the sign. For negative feedback, therefore, we can take the feedback equation to be

$$A_v' = \frac{A_v}{1 + \beta A_v}$$

GAIN STABILITY

As we have seen amplifier gain can be affected by changes in the supply voltage, the ageing of components (particularly valves), and by the replacement of active devices (particularly transistors) with other which do not have the same characteristics as the originals. Negative feedback can do much to counteract the effects of all these variations. An example will be a useful introduction to the case of gain stability.

Example (2) An amplifier with a voltage gain of 20 000 is used in a feedback circuit where $\beta = 0.02$. Calculate the overall gain.

If the gain dropped by one half of its inherent value because of supply voltage variation, what would the overall circuit gain now become?

With the full gain available

$$A_v' = \frac{20\,000}{1 + (0.02 \times 20\,000)}$$

$$= 49.9$$

When the gain falls to 10 000, the new gain with feedback

$$A_v' = \frac{10\,000}{1 + (0.02 \times 10\,000)}$$

$$= 49.75$$

This example illustrates a very important point: although the inherent gain of the amplifier without feedback fell from 20 000 to 10 000, a change of 50%, the overall gain of the system with feedback dropped only from 49.9 to 49.75, a fall of less than 0.5%. The circuit with feedback has therefore a more stable gain figure than the amplifier alone. If we examine the general feedback equation

$$A_v' = \frac{A_v}{1 + \beta A_v}$$

then when A_v is very much greater than 1

$$A_v' \simeq \frac{A_v}{\beta A_v} \simeq \frac{1}{\beta}$$

This makes A_v' independent of A_v, since only β is involved in this expression. Hence the gain of a feedback amplifier is very stable.

Since β is fractional, the way to make βA_v large is to make A_v large by starting off with a very high gain amplifier. This is not always an easy option, as high gain tends to introduce instability because of undesirable positive feedback occurring between the circuit wires and component fields. Negative feedback will do nothing to put right a poor initial design.

(3) An amplifier has a gain of -1500 and is used with a feedback network where $\beta = 0.1$. Calculate the overall voltage gain.

(4) An amplifier has a gain of 10^3 without feedback. Calculate the gain when 0.9% of negative feedback is applied.

(5) You require an amplifier with a final gain of 75. What should be the open-loop gain of the basic amplifier if the feedback fraction is to be -0.01?

(6) An amplifier having an open-loop gain of 500 has overall n.f.b. applied which reduces the gain to 100. What is the feedback fraction?

(7) What would the gain of an amplifier become if the whole of the output voltage was fed back in opposition to the input? Is such a design possible?

ADVANTAGES OF NEGATIVE FEEDBACK

We will now work through a number of advantages that result from the application of n.f.b. to an amplifier.

1. Reduction of Distortion

Any distortion which may be present in the *original* signal input to an amplifier is indistinguishable from the signal and is therefore unaffected by n.f.b. Regarding distortion arising *within* the amplifier, for example, a harmonic introduced in the output stage, if a portion of this is fed back in opposition to the generated harmonic, the amount of distortion will be reduced.

Let an amplifier without feedback be considered as having harmonic distortion introduced by means of a hypothetical voltage generator as shown in *Figure 5.4(a)*. The distortion fraction is then D/V_o where D is the distortion voltage magnitude. Now suppose a feedback network is added to the amplifier as shown in *Figure 5.4(b)*. Since the distortion

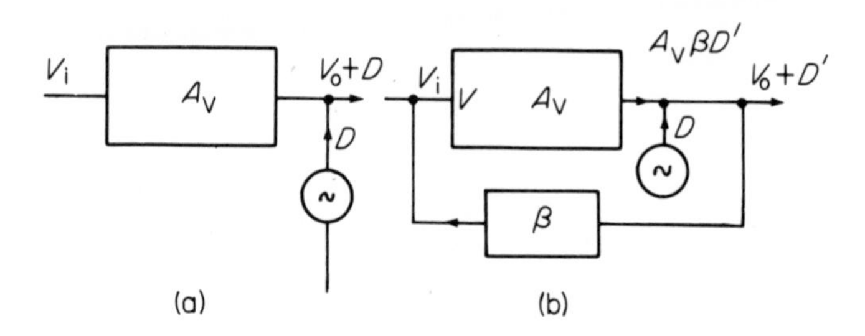

Figure 5.4

usually depends upon the magnitude of the input voltage v, suppose v_i to be so adjusted that the input voltage at the amplifier terminals is the same as before. Then the harmonic voltage fed back = $\beta D'$, say, and this appears at the output as $A_v\beta D'$.

$$\therefore \quad D'(1 - \beta A_v) = D$$

Since

$$A_v\beta D' + D = D'$$

Hence

$$D' = D + A_v\beta D'$$

$$\therefore \quad \frac{D'}{D} = \frac{1}{1 - \beta A_v}$$

$$= \frac{1}{1 + \beta A_v} \quad \text{for } \beta A_v \text{ negative.}$$

So the distortion fraction is improved by the factor $1 + \beta A_v$.

This argument may seem academic at first, for the distortion is reduced by the same factor as is the gain of the system. However, harmonic distortion usually results from large signals traversing the extremes of the dynamic output characteristics and such excursions most often occur in the power output stages of amplifiers. By using n.f.b. over such stages to reduce the distortion and recovering the lost gain in the small signal stages earlier in the amplifier is a common practice. You will probably see a 'catch' at this point, and a logical question to ask is, what happens to the harmonic distortion when we increase the gain to what it was originally? It might be best to answer this by way of a worked example.

Example (8) An amplifier has an open-loop gain of 100 and is used with a feedback network where $\beta = -0.1$. If the fundamental output voltage is 10 V with 20% second-harmonic distortion, calculate (a) the input required to maintain the output at 10 V, (b) the value of second-harmonic voltage when f.b. is applied, (c) the total overall distortion when the gain is restored to its original value.

(a) with feedback $A_v = \dfrac{100}{1 + (0.1 \times 100)} = \dfrac{100}{11} = 9.1$

Hence to provide 10 V output with this gain, the required input will be 10/9.1 = 1.1 V.

(b) Since there is 20% distortion, the distortion voltage at the output without f.b. is 20% of 10 V = 2 V.

Then with f.b. the distortion voltage is reduced to

$$D' = \frac{D}{1 + \beta A_v} = \frac{2}{11} = 0.182 \text{ V}$$

Notice that the gain and the distortion have both been reduced by the same factor i.e. 1/11.

(c) We now want to restore the gain to its original value of 100. Since the gain with f.b. is 9.1, we shall need *three* such stages to obtain a gain of at *least* this figure.

In the original amplifier the percentage harmonic was 20% and with f.b. this drops to 0.182/10 × 100% = 1.82%. The first stage will now introduce 1.82% harmonic distortion and this will be fed as input to the second stage. This will add a further 1.82% harmonic of its own and so the total harmonic will be 3.64% at the output of the second stage. (Make certain you understand why we *add* the distortion percentages). In the same way the third stage will contribute a further 1.82% and the total distortion at the output will be 5.46%.

This is a considerable improvement on the original 20%. Actually, with three stages we have more gain than we need; the signal level in the first two stages could be kept low and the overall harmonic content at a gain of 100 would be much less than the figure of 5.46% calculated.

2. Noise Consider the diagram of *Figure 5.5* which is the same as that shown in *Figure 5.1(b)* but with an additional input signal N which can be assumed to be a noise voltage introduced into the first stage of the amplifier. Then the input signal becomes

$$V = (V_i + N) + \beta v_o$$

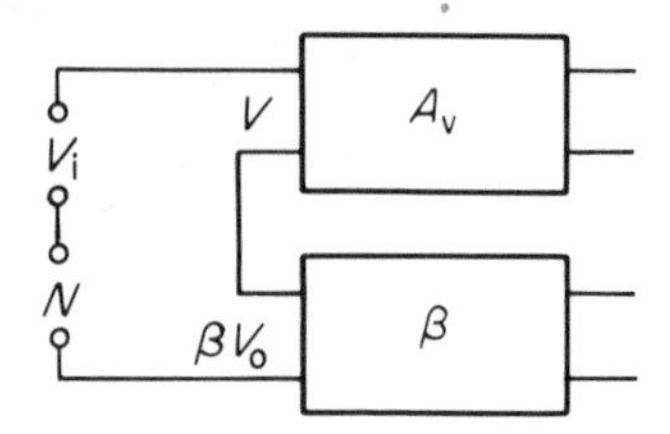

Figure 5.5

and this equation indicates that the noise signal will be dealt with in the same manner as V_i, that is, N will be reduced by the same amount as V_i. Hence

$$N' = \frac{N}{1 + \beta A_v}$$

Noise may be reduced by the application of n.f.b. but because of the requirement for very high forward gain and so an increased number of stages of amplification, it is possible for the overall noise level to be increased. Once again, n.f.b. is not a panacea for a poor original design.

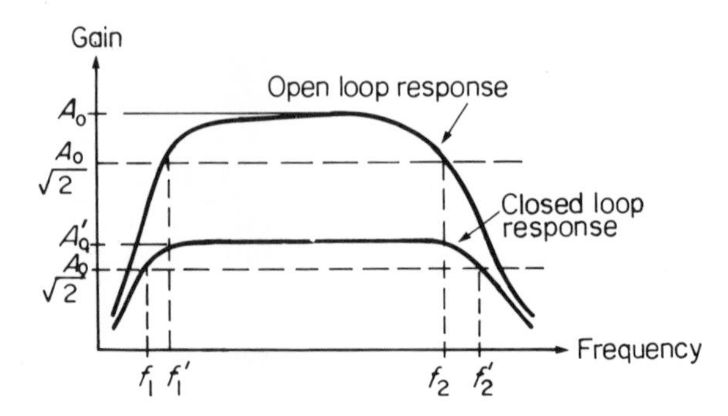

Figure 5.6

(9) Can n.f.b. have any effect on signal-to-noise ratio?

3. Increase in bandwidth

For large amounts of n.f.b. $A_v' \simeq 1/\beta$, hence if β can be made independent of frequency, the overall gain will be independent of frequency. It is not possible to obtain an infinite bandwidth by this method (or any other), but n.f.b. can nevertheless extend the bandwidth of an amplifier by a considerable amount.

As we have already noted, bandwidth is defined as that range of frequencies over which the gain does not fall below $1/\sqrt{2}$ of its maximum (usually its mid-band) value. *Figure 5.6* shows the open-loop response of an amplifier with a bandwidth $f_2 - f_1$ Hz and mid-frequency gain A_o. When n.f.b. is applied this gain is reduced to $A_o' = A_o/(1 + \beta A_o)$. A study of the diagram makes it clear that the bandwidth has now increased to $f_2' - f_1'$. It can be proved that

$$f_2' = f_2(1 + \beta A_o)$$

and

$$f_1' = \frac{f_1'}{1 + \beta A_o}$$

As f_1' is often reduced to a very low frequency, to a good approximation the overall bandwidth with large amounts of n.f.b. is f_2'.

A figure of merit for a wideband amplifier is its gain-bandwidth product. Negative feedback reduces the gain but increases the bandwidth; overall the figure of merit may be relatively unaffected.

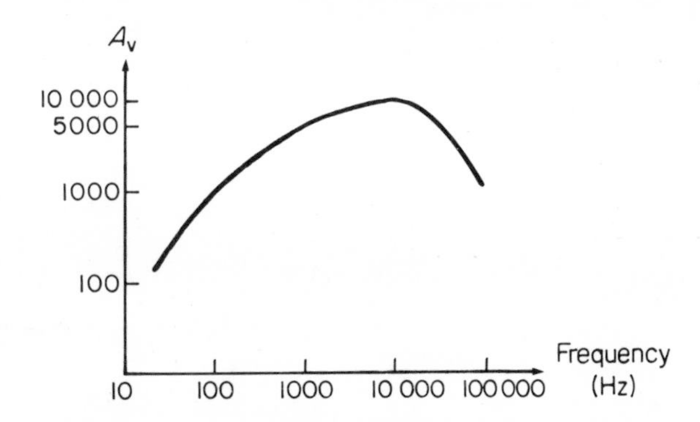

Figure 5.7

(10) The response curve of an amplifier is given in *Figure 5.7.* Calculate the resultant values of gain at 100 Hz, 1 kHz, 10 kHz and 100 kHz when 1% of the output voltage is fed back in antiphase to the input. Sketch (on the same axes) the resultant response curve.

CURRENT FEEDBACK PRINCIPLES

Before continuing the list of advantages arising from the use of n.f.b. we will consider the cases of feedback amplifiers where the signal fed back is applied in *parallel* with the input terminals. In these cases we are concerned with the addition of *currents*, not voltages, and so these forms of feedback are given the names *current-current* f.b. or *current-voltage* f.b., depending upon the manner in which the signal is derived at the output end.

Figure 5.8(a) and (b) show respectively current-current f.b. and current-voltage f.b. By simple consideration, parallel connected feedback will *not* affect the voltage gain of the amplifier since for a given voltage output the *same* input voltage is required whether or not the

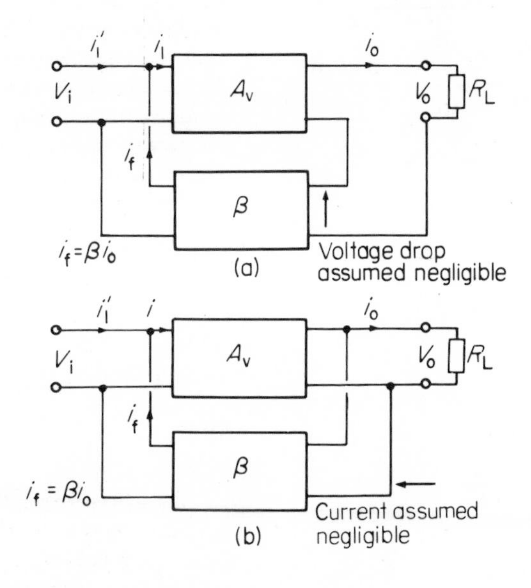

Figure 5.8

feedback circuit is connected. The current gain is, however, affected, for from *Figure 5.8(a)*

$$i_f = i_o$$

$$i_1 = i_1' + i_f$$

$$i_1' = \frac{i_1}{1 + \beta A_i}$$

where A_i is the current gain ratio i_o/i_1

$$A_i' = \frac{A_i}{1 + \beta A_i}$$

taking βA_i as being negative. This is analogous to the series voltage feedback cases with current gain A_i replacing voltage gain, and being the current ratio i_f/i_o.

This relationship is not of such interest to us as most amplifiers are designed as voltage amplifiers, and for those using FET or valve input stages, the current amplification has little meaning.

We now investigate the effect of n.f.b. on the input and output resistance of amplifiers.

4. Input resistance

The input resistance of an amplifier is the resistance as measured at the input terminals. In the following paragraphs we shall consider only the case of amplifiers which, before feedback is applied, have purely resistive input and output impedances. For most amplifiers, this is a reasonably true state of affairs.

There are two cases to consider from the point of view of the effect of n.f.b. on input resistance: (a) feedback is in series, (b) feedback is in parallel with the input.

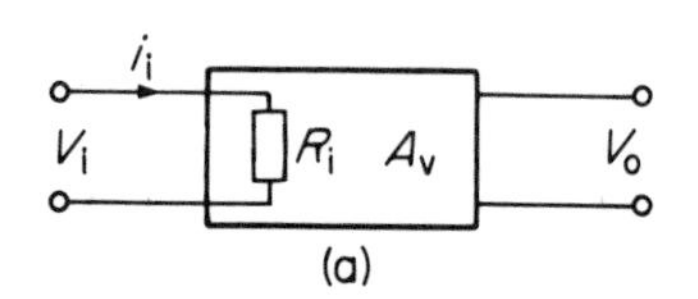

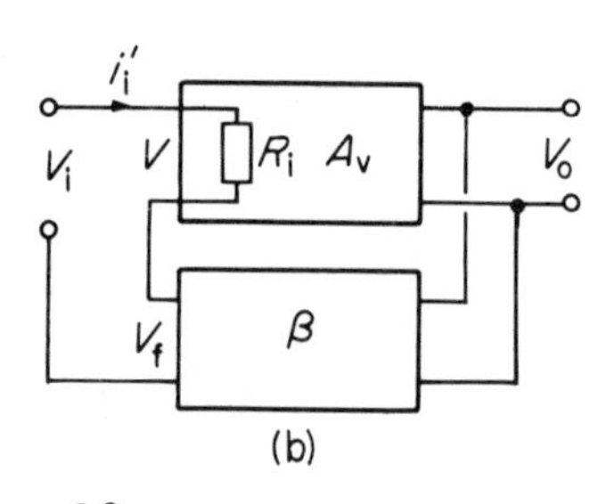

Figure 5.9

(a) *Figure 5.9(a)* shows an amplifier without feedback, of open-loop gain A_v and input resistance R_i. This resistance can be measured as being simply the ratio of V_i to i_i. In diagram (b) feedback has been applied in series with the input. Now since V_f is opposing V_i, it is evident that the effective voltage V at the input terminals will now be less than V_i and hence, as seen from these terminals, the input current will be *reduced* from i_i to i_i'. This means that the input resistance as seen from the *source* terminals has apparently *increased.* Let us check this out with a little algebra.

The input resistance with feedback will be

$$R_i = \frac{V_i}{i_i}$$

and

$$V_i + V_f = i_i' R_i \qquad (5.2)$$

Now

$$V_o = A_v'.V_i = V_i \frac{A_v}{1 + \beta A_v}$$

$$\therefore \quad V_f = \frac{\beta A_v.V_i}{1 + \beta A_v} \quad \text{since } V_f = \beta V_o$$

Substituting this value of V_f into (5.2):

$$V_i + \frac{\beta A_v . V_i}{1 - \beta A_v} = i_i' R_i$$

$$\therefore \quad V_i \left[1 + \frac{\beta A_v}{1 + \beta A_v}\right] = i_i' R_i$$

But with feedback

$$R_i' = \frac{V_i}{i_i'} = R_i(1 + \beta A_v)$$

Hence, as we deduced, the input resistance is increased with series applied n.f.b. by a factor $(1 + \beta A_v)$.

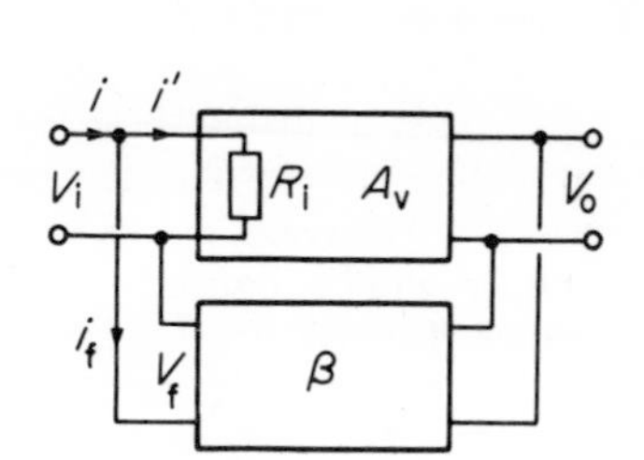

Figure 5.10

(b) We now consider the situation where the feedback is applied in parallel with the input as shown in *Figure 5.10*. In the parallel circuit, as mentioned, we are concerned with current addition, not voltage. The input current i is the sum of the amplifier input current i' and the feedback current i_f. So the input resistance R_i' with feedback applied is

$$R_i' = \frac{V_i}{i' + i_f} \text{ where } i_f = \frac{V_f}{R_i} \text{ and } i' = \frac{V_i}{R_i}$$

$$\therefore \quad R_i' = \frac{V_i}{\frac{V_i}{R_i} + \frac{V_f}{R_i}} = \frac{V_i R_i}{V_i + \beta A_v . V_i}$$

$$R_i' = \frac{R_i}{1 + \beta A_v}$$

So, with parallel connected feedback the input resistance is *reduced* by a factor $(1 + \beta A_v)$.

Example (11). An amplifier has an open-loop gain of 5×10^5 and an input resistance of 100 kΩ. When n.f.b. is applied in series with the input, the gain is reduced to 10^3. Find (a) the value of the feedback fraction β, (b) the new input resistance.

(a) $$A_v' = \frac{A_v}{1 + \beta A_v}$$

$$\therefore \quad A_v = A_v' + \beta A_v A_v'$$

$$\therefore \quad \beta = \frac{A_v - A_v'}{A_v A_v'} = \frac{(5 \times 10^5) - 10^3}{5 \times 10_{\cdot}}$$

$$= 0.998 \times 10^{-3}$$

(b) For series connection $R_i' = R_i(1 + \beta A_v)$

$$= 100 \times 10^3 (1 + [0.998 \times 10^{-3} \times 5 \times 10^5])\Omega$$

$$= 100 \times 10^3 (1 + [4.99 \times 10^2])\Omega$$

$$= 100 \times 10^3 \times 500\Omega$$

$$= 50 \text{ M}\Omega$$

5. Output resistance

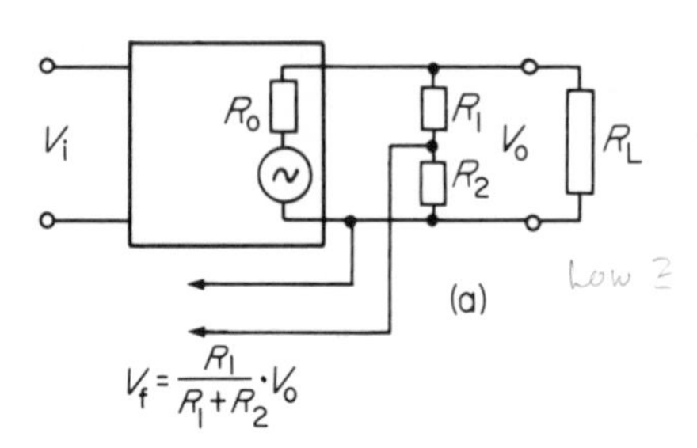

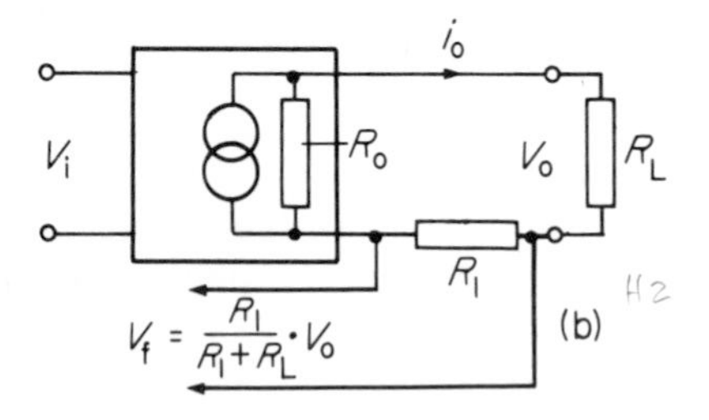

Figure 5.11

The output resistance of an amplifier is the resistance measured at the output terminals when the input is zero. There are again two cases to consider when feedback is taken from the output terminals, and *Figures 5.11(a) and (b)* show the possible arrangements. Diagram (*a*) illustrates the feedback network connected in parallel with the output, the method we mentioned at the beginning of this section. This is the case of voltage feedback because V_f is directly proportional to V_o and β the feedback fraction is given as $R_2/(R_1 + R_2)$. It is assumed that the total resistance of R_1 and R_2 in series is very much greater than the load R_L so that the introduction of the feedback network does not affect the loading.

Diagram (*b*) illustrates current feedback; here the load current is passed through a resistor R_1 in series with load R_L. The feedback quantity is now proportional to the output current i_o and the feedback fraction β is $R_1/(R_1 + R_L)$. R_1 is very small compared with R_L so that the load current is unaffected by its introduction. β then approximates to R_1/R_L.

The change in output impedance with f.b. may be deduced qualitatively by considering the diagrams in turn. Suppose in (*a*) V_o increases for some reason but V_i remains constant. V_f will increase, but as the feedback is negative the effective input voltage will be reduced and this in turn will tend to reduce the output voltage. The net effect is to keep V_o constant, hence the output circuit acts as a device with a *small* internal resistance i.e. a constant-voltage source. The diagram illustrates this form of output equivalent circuit.

Conversely, in diagram (*b*), if i_o increases, V_f increases and the effective input quantity (current or voltage) will decrease. The net effect is to keep i_o constant and so the output circuit behaves this time as a constant-current generator i.e. a device with a *large* output resistance. This is shown on the diagram.

Hence we deduce that voltage-derived (parallel) feedback reduces the output resistance and current-derived (series) feedback increases the output resistance. You must not assume that these changes in output resistance come about because the parallel feedback resistors are shunting the load, or that the series feedback resistor adds to the load. We have already stated that the parallel resistor draws negligible current and the series resistor drops negligible voltage.

It can be proved that for the parallel connection

$$R_o' = \frac{R_o}{1 + \beta A_v}$$

and for the series connection

$$R_o' = R_o(1 + \beta A_v)$$

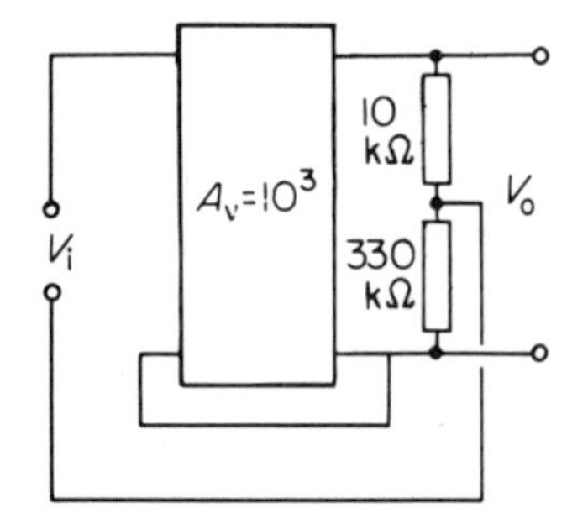

Figure 5.12

Example (12) In the feedback circuit shown in *Figure 5.12*, the amplifier gain is 10^3 when the 330 Ω resistor is short-circuited. What is the gain when the short-circuit is removed?

If the resistors used have a tolerance of ±5%, calculate the maximum and minimum values that the gain might have with feedback.

When the 330 Ω resistor is short-circuited, there is no feedback and so the gain $A_v = 1000$.

When the 330 Ω resistor is in circuit, feedback is applied and the feedback fraction $\beta = 0.33/(10 + 0.33) = 0.032$.

$$\therefore \quad A_v' = \frac{A_v}{1 + \beta A_v} = \frac{1000}{1 + (0.032 \times 1000)} = 30.3$$

If the resistors are ±5% tolerance, we must find the extreme values that β can take. Assuming that the 330 Ω resistor is at the top end of its tolerance range and the 10 kΩ resistor is at the bottom end of its range, their respective values are 346.5 Ω and 9.5 kΩ. Then

$$\beta = \frac{0.3465}{9.5 + 0.3465} = 0.0352$$

and

$$A_v' = \frac{1000}{1 + 35.2} = 27.6$$

When the 330 Ω resistor is at the bottom end of its tolerance range and the 10 kΩ resistor is at the top end, their respective values are 313.5 Ω and 10.5 kΩ. Then

$$\beta = \frac{0.3135}{10.5 + 0.3135} = 0.029$$

and

$$A_v' = \frac{1000}{1 + 29} = 33.34$$

The gain could therefore lie between the limits 27.6 to 33.34.

Try the next two examples on your own.

(13) If the gain of an amplifier without feedback is 55 dB, what will be the resultant gain in dB when negative feedback is applied from the output to the input and $\beta = 0.1$?

(14) An amplifier has an open-loop gain of 300 which is found to fall by 20% due to changes in the supply voltage. If the gain is to be stabilized so that it falls by only 1%, calculate the required value of feedback fraction β. If the original bandwidth of this amplifier was 500 Hz to 50 kHz, calculate the bandwidth when feedback is applied.

INSTABILITY IN NEGATIVE FEEDBACK AMPLIFIERS

On several occasions throughout this section mention has been made of the necessity of careful design when negative feedback is being applied to an amplifier system. It is often put about that negative feedback is used to prevent oscillation in the amplifier, but this is quite wrong. Indeed, negative feedback carelessly applied will often cause a perfectly stable amplifier to become unstable.

One of the major problems associated with negative feedback is that of phase shift around the loop. At the so-called mid-frequencies of the amplifier passband, approximating to the normal 3 dB bandwidth range, there is a 180° phase shift in each stage of a voltage amplifier, and the feedback voltage is arranged to be in phase opposition to the input signal. At frequencies beyond the passband, the reactive components within the amplifier introduce additional phase shifts, so that the total phase shift moves from 180° down towards 90° or up towards 270°. It is then possible for the feedback to become positive at certain low or high frequencies, the feedback signal adding to the input instead of opposing it. The result is that the circuit will become unstable and oscillate, so invalidating its function as an amplifier (or at least as an amplifier whose parameters are under our control). This trouble can only be eliminated by ensuring that the loop gain βA_v is less than unity at the frequencies in question.

The emitter follower

The common-collector circuit of *Figure 5.13*, or as it is more usually known, the *emitter-follower* amplifier, is an example of voltage-voltage feedback. The load R_L is in the emitter circuit, resistors R_1 and R_2 are the usual bias components, and the collector is returned directly to the V_{cc} line where, from the signal point of view, it is effectively at earth potential. Assume that we apply a sinusoidal signal at the input terminals.

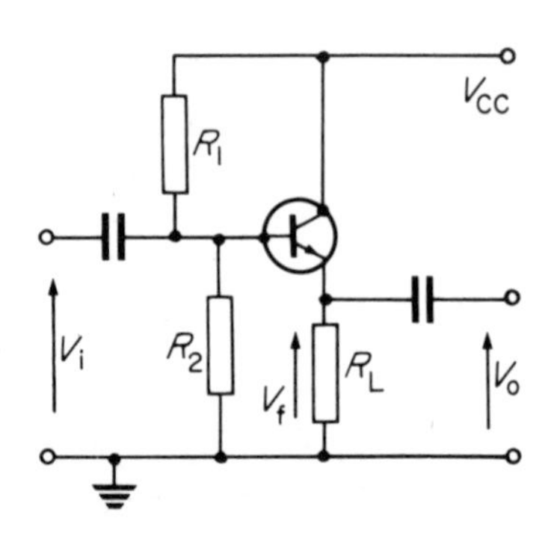

Figure 5.13

As V_i rises towards a maximum, emitter current will increase and so the emitter voltage will rise relative to earth. The converse will happen when the input goes negative. Hence the emitter voltage "follows" the input voltage and V_i and V_o are *in phase relative to the earth line.* However, in the base circuit input *relative to the emitter,* V_i and V_f are in series, hence the effective base-emitter potential is the difference between V_i and V_f and these voltages are therefore in *phase opposition*, see *Figure 5.14.* Notice that in this circuit V_o is in phase with V_i and hence $A_v = V_o/V_i$ is positive, but V_f is phase opposed to V_i and hence to V_o, therefore $\beta = V_f/V_o$ is negative. But $V_f = V_o$, hence $\beta = -1$. This means that *all* the output voltage is fed back in series with the input, and 100% negative feedback is applied.

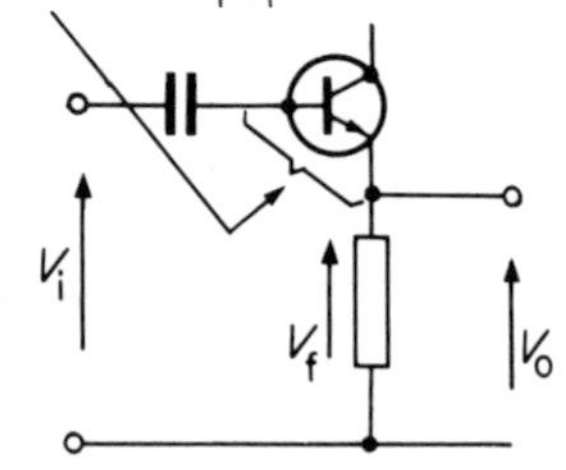

Figure 5.14

Substituting $\beta = -1$ in the feedback equation (5.1) we get

$$A_v' = \frac{A_v}{1 - \beta A_v} = \frac{A_v}{1 + A_v}$$

This must be less than unity; hence the gain of the emitter-follower is always slightly less than 1 and there is no voltage phase reversal.

The usefulness of the emitter-follower stems from the effect that 100% n.f.b. has on the input and output resistances. As the circuit is voltage-voltage feedback, with the feedback derived in parallel with the output load and fed back in series with the input circuit, the respective resistances can be expected to be high at the input and low at the output terminals. Hence the amplifier is ideal as an 'electronic transformer' for matching a stage of high resistance output into one of low resistance input without voltage reduction or frequency restriction inherent in an ordinary step-down transformer. The low output resistance is also useful when pulse signals are fed to a capacitive device such as a cathode-ray tube, distortion of the pulses being greatly reduced.

It can be proved that to a good approximation

$$R_i' \simeq h_{fe}R_L$$

(but account has to be taken of the shunting effect of the bias resistors), and

$$R_o' \simeq \frac{R_S}{h_{fe}}$$

where R_S is the resistance of the signal source feeding the amplifier. This value of R_o' has R_L effectively in parallel with it.

Bootstrapping

It is the shunting effect of the bias resistor(s) in the emitter-follower that degrades the very high input resistance that this amplifier otherwise provides. By arranging the circuit in the manner shown in *Figure 5.15* the paralleling effect of bias resistors R_1 and R_2 upon the input is practically eliminated. This is known as *bootstrapping*.

Capacitor C acts as a d.c. block between the emitter and base circuits but is selected to have a negligible reactance at the signal frequency. Hence feedback occurs between the output signal at the emitter terminal and the input signal at the base through this capacitor. Because $A_v \simeq 1$, the signal amplitude across R_L is almost of the same amplitude as the incoming signal, in addition it is in phase with the incoming signal. Hence the potentials at any instant at either end of resistor R_3 are approximately equal. This is equivalent to R_3 having a very high resistance to the *signal frequency* but *not* to the d.c. bias current feeding the base. Hence the shunting effect of R_1 and R_2 is isolated from the input terminals.

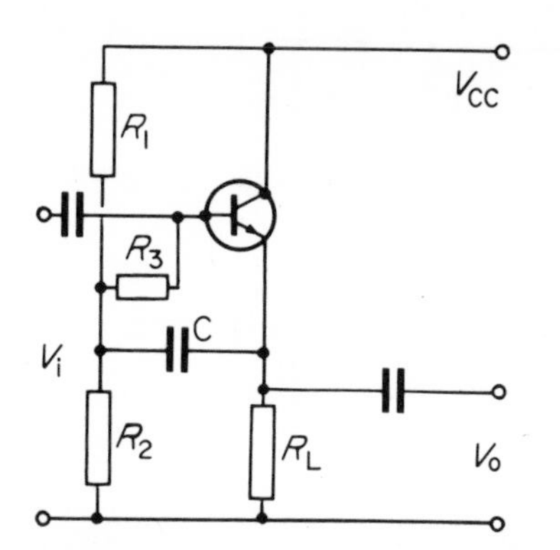

Figure 5.15

(15) Would it help if the value of R_3 in the bootstrap circuit was made very large, say, several megohms?

SOME PRACTICAL FEEDBACK CIRCUITS

Current-voltage feedback

Figure 5.16 illustrates a simple example of current-voltage feedback. You will probably recognise this circuit as being one of the examples mentioned in Section 2 when bias point stability was being discussed. By returning the bias-resistor R_2 to the collector of the transistor instead of to the V_{cc} line, any tendency for I_C to increase is countered by a reduction in base current, and conversely, so that the base bias is held reasonably stable against temperature and supply voltage fluctuations.

Well, we can now look at the same circuit from the point of view of a negative-feedback amplifier, where a current whose magnitude is proportional to the output voltage is fed back in antiphase to the input current signal. If you glance back at *Figure 5.8(b)* you will recognise this form of feedback. Provided that the feedback resistance R_1 is large compared to the input resistance of the amplifier, the feedback current $\beta i_o \simeq V_o/R_1$. The output current i_o is equal to V_o/R_2. hence

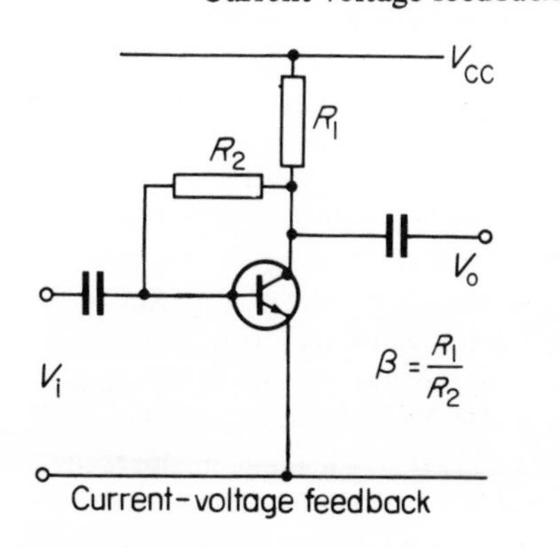

Figure 5.16

$$\frac{\beta i_o}{i_o} = \frac{V_o/R_1}{V_o/R_2}$$

and so

$$\beta = \frac{R_1}{R_2}$$

In this circuit R_1 has not only to do its job as feedback resistor, it also provides the d.c. bias for the amplifier.

Voltage-voltage feedback

Figure 5.17 illustrates an example of voltage-voltage feedback, a two-stage amplifier being concerned. The feedback path is from the collector output of the second transistor via resistance R_2 back to the emitter of the first. R_1 and R_2 are therefore effectively in series across the amplifier output terminals, and that part of the output voltage developed

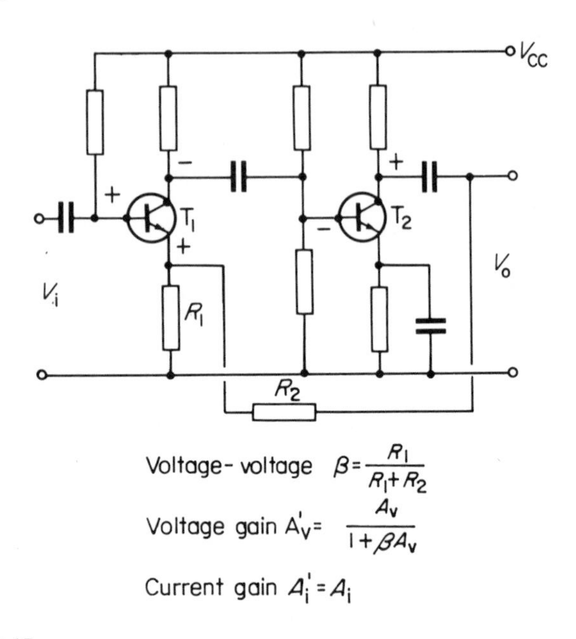

Figure 5.17

across R_1 is fed back in phase opposition to the input signal voltage. A glance back to *Figure 5.11(a)* will show you this form of feedback in its basic block.

The signs indicated in *Figure 5.17* show that the signal fed back to the emitter has the *same* sign as the signal voltage on the base. Hence, in the same way as feedback occurs in the common emitter circuit, the effective base-emitter input voltage is *reduced* by the feedback.

If you are thinking in terms of phase shift round the loop you might see a snag here. The phase shift through the amplifier itself is zero (each transistor introduces 180°), and clearly there will be no phase shift in the resistive feedback network; so the *total* loop shift is zero. Such a situation leads to positive feedback and instability and this is due to the way in which the feedback is introduced into the emitter of the first transistor. It is true that the feedback signal arrives back with the same phase as the base signal, but by connecting it to the emitter, the feedback terminals are 'reversed' as it were and the required antiphase condition occurs in the base-emitter circuit of the first transistor. If R_2 were returned to the *base* of this transistor it would be a very different story!

Current-current Feedback

A two-stage example of current-current feedback is shown in *Figure 5.18*, and this was illustrated in block form in the earlier *Figure 5.8(a)*. Here the feedback resistors concerned are R_1 and R_2. The output current i_o of the second transistor splits into two parts at the emitter terminal: one part flows through R_1 to earth and the other part flows

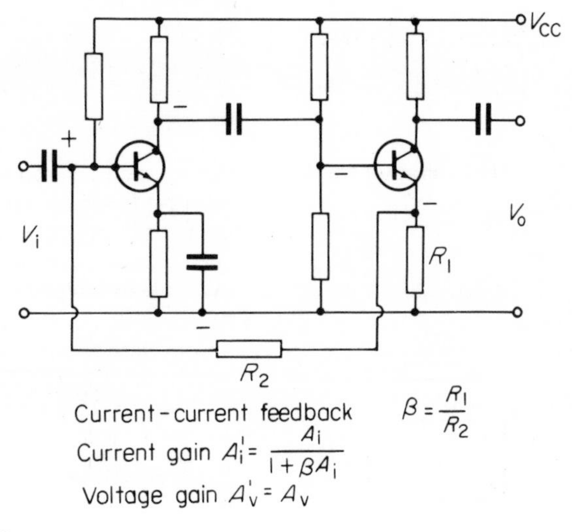

Figure 5.18

through R_2 and the input resistance of the first transistor to earth. If we assume that R_2 is very large compared with the input resistance (which is easily realised), the feedback current βi_o is

$$\beta i_o = \frac{i_o R_1}{R_1 + R_2}$$

and so

$$\beta = \frac{R_1}{R_1 + R_2} \simeq \frac{R_1}{R_2} \text{ if } R_2 \gg R_1$$

Compare the circuit with the previous one and follow through the phase conditions, again indicated by the signs. Although the amplifier is two-stage, the phase shift in it (for the feedback) is only 180°. The second transistor introduces no phase shift between its base and emitter from which the feedback is derived. So this time the feedback is applied to the base of the first transistor where it is clearly antiphase to the input current.

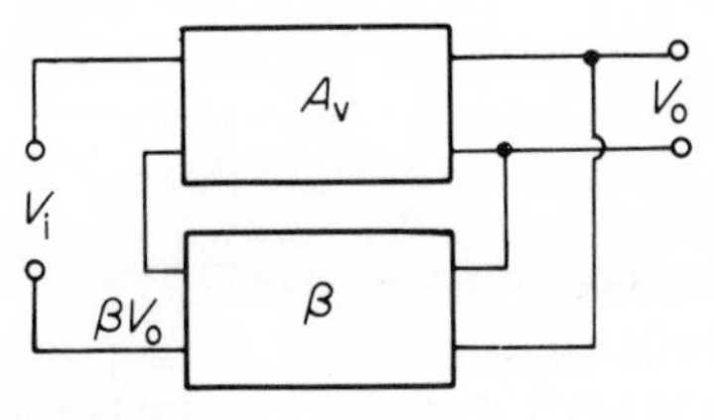

Figure 5.19

(16) In the circuit of *Figure 5.17*, there is additional feedback being applied to the first transistor besides that deriving from the output via R_2. Can you explain where and why this is happening?

(17) With reference to the block diagram of *Figure 5.19*, fill in blank spaces in the following table:

A_v	β	βA_v	A_v'
100	–0.1		
1500			48.4
	–0.05		17.7
25		–6.25	

The operation amplifier

An operational amplifier is a very high gain voltage amplifier with negative feedback applied between input and output, and is used in computer applications, as its name implies, to carry out a number of specialised mathematical operations. These amplifiers are readily available in integrated circuit form, and the type and amount of feedback can be selected to suit any particular application by connecting appropriate components between its terminals. We shall consider it here purely as a form of negative-feedback amplifier.

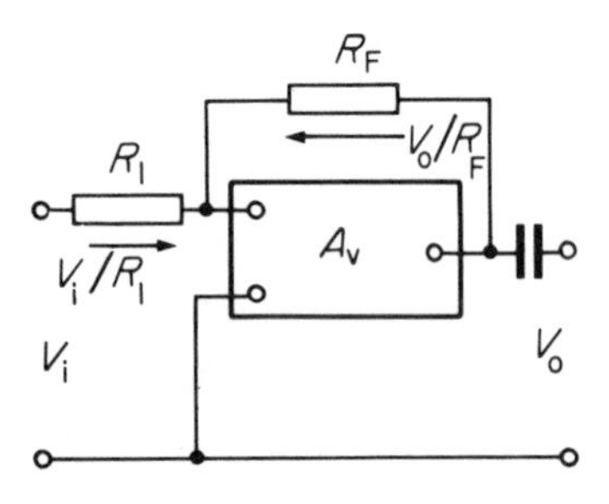

Figure 5.20

Figure 5.20 shows a block diagram of an operational amplifier using a typical form of n.f.b. We assume here that there is a 180° phase shift in the amplifier, i.e. it is connected as an inverting amplifier. A resistor R_1 is connected in series with the signal input lead and a feedback resistor R_f is connected between the input and output terminals of the amplifier. As a result, the large n.f.b. voltage which opposes the input signal, reduces the actual input to the amplifier to a very small figure.

The current provided by the input is V_i/R_1 and practically the whole of this flows into R_f. The current in R_f due to the output voltage V_o is V_o/R_f. By Kirchhoff's current law, the sum of these two currents flowing into the input junction must be zero, hence

$$\frac{V_i}{R_1} + \frac{V_o}{R_f} = 0$$

and so

$$V_o = -\frac{R_f}{R_1} V_i$$

This shows us that the output voltage is simply the input voltage (with a 180° phase change) multiplied by the ratio of the two resistors. Hence

$$A_v' = \frac{V_o}{V_i} = -\frac{R_f}{R_1}$$

So the gain with feedback depends only upon the ratio of the resistors and is *independent* of the open-loop gain of the amplifier, providing that this is very high. In practical operational amplifiers, the open-loop gain at low frequencies can be more than 100 000.

This is as far as we need to go in the study of negative feedback at the present stage of the course. You should examine commercial amplifier circuits, locate the feedback loops, and decide for yourself what forms of feedback they are. This way you will become acquainted with the many circuit arrangements that are possible in the application of negative feedback to amplifiers.

PROBLEMS FOR SECTION 5

(18) Define the terms: (a) open-loop gain, (b) feedback fraction, (c) gain stability.

(19) Say whether each of the following statements are true or false:

Series connected n.f.b. (a) increases the input resistance.
(b) reduces the voltage gain.
(c) increases the current gain.

Parallel connected n.f.b. (a) increases the output resistance.
(b) leaves the voltage gain unaffected.
(c) reduces the current gain.

(20) The voltage gain of an amplifier is A. If a fraction β of the output voltage is fed back in antiphase with the input, derive an expression for the new overall gain of the amplifier.

(21) In a certain amplifier $A_v = 50$. Find the value of A_v' if (a) $\beta = 0.005$; (b) $\beta = 0.02$; (c) $\beta = -0.02$. Comment on each of your solutions.

(22) A transistor provides a gain of 50. If feedback is applied so that the gain increases to 62.5, what sort of feedback is it and what fraction of the output is fed back?

(23) An amplifier has 0.5% of its output voltage fed back in antiphase to the input. If the gain of the amplifier is then 160, what was its gain before feedback was applied?

(24) An amplifier has an open loop gain of 1000 which is likely to vary by 10% because of supply fluctuations. If feedback with $\beta = -0.01$ is applied to the amplifier, calculate the overall variation in gain.

(25) An amplifier with a voltage gain of 2×10^3 is used in a feedback arrangement where $\beta = -0.02$. Calculate the overall gain. If the gain dropped to one-half of its inherent value, what would the overall circuit gain become?

(26) An amplifier has a voltage gain of 57 dB. When n.f.b. is applied the gain falls to 27 dB. Calculate β.

(27) An amplifier has a voltage gain of 60 dB without feedback and 30 dB when feedback is applied. If the gain without feedback changes to 55 dB, calculate the new gain with feedback.

(28) Negative series feedback is applied to an amplifier to reduce its open-loop gain by 50%. By what factor will the input and output resistances change?

(29) An amplifier has a voltage gain of 60 dB. Find the dB change in the gain if 1/30 of the output signal is fed back to the input in phase opposition. Ignoring any phase change in either the amplifier or the feedback network, calculate the reduction in harmonic distortion at the amplifier output.

(30) An amplifier circuit is to be designed to have an overall gain of $20 \pm 0.1\%$. The basic amplifier from which the final circuit is to be derived has a gain $A_v \pm 10\%$. Calculate the required value of β and the gain A_v.

(31) A n.f.b. amplifier is shown in *Figure 5.21*. When the 1 kΩ resistor is short-circuited the voltage ratio V_o/V_i is 2000. What will this ratio be when the short-circuit is removed?

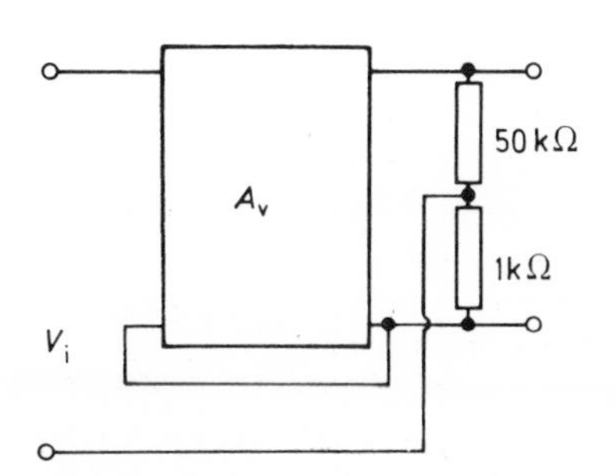

Figure 5.21

(32) An amplifier has a mid-frequency voltage gain of 400. A circuit giving 5% n.f.b. is added to it. Calculate:

(a) the new overall mid-frequency gain:

(b) the new gain at those frequencies where the gain of the basic amplifier has fallen to the 3 dB points:

(c) the factor by which the overall gain has been reduced at the 3 dB frequencies.

The original bandwidth of this amplifier extended from 100 Hz to 10 kHz. What will the bandwidth limits be with n.f.b. applied?

(33) An amplifier has an open-loop gain of 500, a bandwidth of 100 kHz and an imput resistance of 1 kΩ. By the use of n.f.b. applied in series with the input the bandwidth is increased to 5 MHz. What will be the new values of gain and input resistance?

(34) A FET has a mutual conductance of 1.5 mS and is employed as a source-follower with a source load of 2.7 kΩ. If this circuit feeds into a device whose input resistance is 4.7 kΩ, what will be the voltage gain of the circuit?

(35) What is the value of β in an emitter-follower amplifier? Show that the voltage gain of such an amplifier is approximately equal to unity.

(36) Find the approximate power gain (in dB) of an emitter-follower having an input resistance of 100 kΩ and an emitter load of 1 kΩ.

(37) In the absence of negative feedback the voltage gain of a wideband amplifier varies with frequency as follows:

Frequency (kHz)	1	10	100	1000
Gain (dB)	25	35	40	30

Draw a curve showing the variation in gain with frequency when negative feedback is applied such that the input to output voltage ratio of the feedback path is 12 dB.

(38) An amplifier without feedback has an input impedance equal to a shunt capacitance of 30 pF. If the open-loop gain is 20 and 10% n.f.b. is added in series with the input, what will be the apparent input capacitance? (Not very difficult but be careful).

(39) A n.f.b. amplifier is found to have a voltage gain of 25. When the feedback leads are interchanged at one end, the gain is 85. Determine (a) the feedback fraction, (b) the amplifier gain without f.b.

6 Oscillators

Aims: At the end of this Unit section you should be able to:
Understand the operation of an oscillator as an amplifier with positive feedback.
Explain the operation of L-C tuned oscillators and R-C oscillators as sinusoidal oscillators.
State the factors that affect frequency stability and describe methods for improving such stability.
Explain the operation of the astable, monostable and bistable multivibrator.

Up to this point our discussion has centred mainly about the use of the transistor as an amplifier. However, the transistor is not restricted in its applications solely to the raising of a signal level and in this section its function as an oscillator will be examined. As an oscillator is fundamentally an amplifier which supplies its own input signal, many of the principles already considered with reference to amplifiers will apply equally well to the understanding of oscillator action.

Oscillators are extremely important in electronic and communication engineering for they provide signal sources for electronic measurements and are generally used as such in combination with other measuring devices under the name of signal generators or function generators. Oscillators operating at fixed frequencies, or operating over only a restricted range of frequencies are also to be found in transmitters and in all radio and television receivers.

An oscillator is essentially a power converter in the sense that its only input is the d.c. power supply and its output is a continuous waveform which may or may not be sinusoidal in shape. We have seen that a voltage amplifier will, in general, convert a small a.c. input signal into a large a.c. output signal. If a part of the output is fed back to the input of the amplifier then the amplifier will be providing its own input. In a circuit diagram of an oscillator, therefore, we will not expect to find a terminal which is specifically the input signal terminal, although there will be an output terminal from which the generated oscillation may be taken off.

There are two basic forms of oscillator:

(i) Those in which the generated waveform is basically sinusoidal; these are known as *sinusoidal* or *harmonic* oscillators and generally take the form of tuned-feedback or negative-resistance oscillators;

(ii) Those in which the generated waveform is markedly non-sinusoidal, being characterized in fact by sudden changes from one condition of circuit stability to another. These oscillators are known as *relaxation* oscillators.

Our concern in this section is with the basic forms and operation of both these types of oscillator.

POSITIVE FEEDBACK

From section 5 which was concerned with feedback applied to amplifiers, we recall the expression for the overall gain A' of an amplifier with feedback:

$$A' = \frac{A}{1 - \beta A}$$

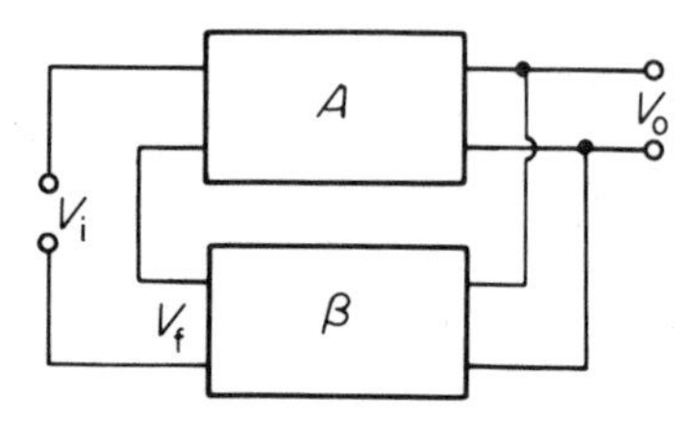

Figure 6.1

where A is the gain of the amplifier without feedback and β is the feedback ratio V_f/V_o. We recall also that mention was made of the particular condition which existed when the denominator of this expression equalled zero, for then the loop gain $\beta A = 1$ and A' became infinite. This condition represents oscillation for it implies that an output signal is available when the external input is zero; and this relates to our description of an oscillator given above, where a continuous signal output is obtained without any form of *external* signal input terminal being apparent.

For convenience, *Figure 5.3* of the previous sectionis brought forwards as *Figure 6.1.* Positive feedback taking place round the amplifier A and feedback network β creates the required situation for continuous oscillation to occur. There are two conditions which must be satisfied for such a system to work and we can consider these in relation to *Figure 6.1.*

(i) The amplifier must be capable of supplying sufficient power to compensate for that which is dissipated in the resistance of the feedback network and still supply enough power to its own input terminals to maintain the required level of output.

(ii) There must be an overall zero or 360° phase shift round the complete loop so that the output signal from the feedback network will be precisely in step with the required amplifier input signal. Thus, if we are using a single stage common-emitter amplifier, for example, which is itself introducing a 180° phase shift between input and output, there must be a further 180° phase shift in the feedback network if oscillation is to be possible.

We can deduce from a consideration of these two requirements that the amplifier circuit forms that part of the feedback loop which maintains the *amplitude* of the output waveform, and it is the β network which determines the *frequency* of the generated waveform. In practice, the loop gain βA must actually be greater than unity, for if it were precisely unity then variations in the supply voltages, ageing of components and possibly temperature or other environmental variations would in all likelihood result in a collapse in the oscillatory output because of a reduction in the overall gain.

When βA is made slightly greater than unity, more signal is fed back than is actually required for oscillation to occur and a build-up in signal level around the loop follows; this build-up cannot continue indefinitely but is quickly limited by non-linearities within the amplifier circuit and by the finite value of the collector supply voltage. Such non-linearities as we have noted earlier will always introduce distortion into the output waveform, but by keeping βA very close to unity such distortion can be held to negligible proportions. In any event, waveshape can be improved (when it is needed) by following circuitry and special amplifiers are often used for this purpose.

(1) For an oscillator to start and operate satisfactorily there must be: (a) a feedback loop in which βA is initially greater than .... (b) a circuit which defines the ....

(2) An oscillator producing a rectangular wave output would be a .... oscillator.

SINUSOIDAL OSCILLATORS

Sinusoidal oscillators may use either tuned inductance-capacitance circuits in the feedback path or combinations of resistance and capacitance, this latter form being known as phase shift oscillators.

In sinusoidal oscillators using coil and capacitor circuits, the frequency of oscillation is fundamentally determined by the LC product, and to a close approximation

$$f = \frac{1}{2\pi\sqrt{LC}}$$

the resonant frequency of a high-Q parallel tuned circuit.

Figure 6.2(a) shows the tuned circuit in its position as the frequency determining unit in the feedback path of the system. Suppose the amplifier A consists of a single stage common-emitter amplifier. The transistor amplifies at the resonant frequency of the tuned circuit and the collector voltage is in phase opposition to the base voltage. A fraction of the collector voltage is taken from a coupling coil L_2, reversed in phase by the appropriate coupling connection and fed, by

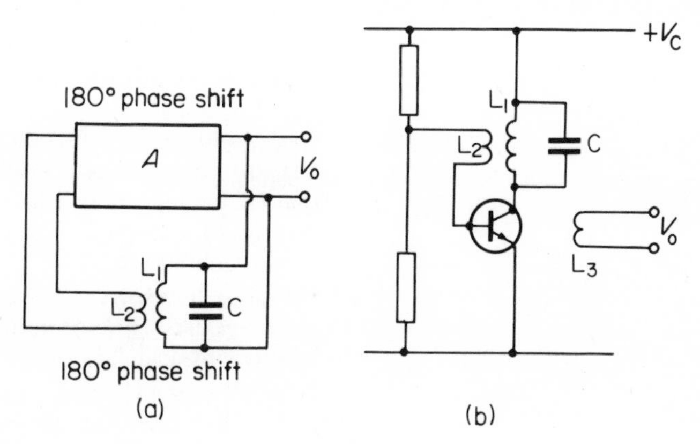

Figure 6.2

mutual coupling, to the amplifier input. The transistor thus provides its own input and the total phase shift is 360°; oscillation is now possible.

Figure 6.2(b) shows the diagram of (a) translated into a practical *tuned-collector* circuit. The collector load consists of the resonant L_1C circuit which is inductively coupled to the base circuit coil L_2. If the voltage induced in the base circuit has the correct phase (depending upon the coil connections) and is of sufficient magnitude (depending upon the degree of coupling M), the system will oscillate. The coil L_3 is used simply to provide a suitable output terminal.

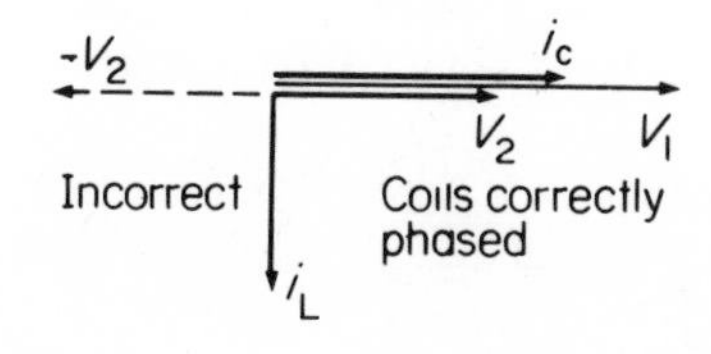

Figure 6.3

The process is as follows. When the collector supply is switched on, collector current commences to flow. The increasing flux in L_1 links with the turns of L_2 and induces an e.m.f. in the base circuit; this e.m.f. will develop between base and emitter as an input signal. A simplified phasor diagram is shown in *Figure 6.3*. Starting with collector current i_c, the voltage V_1 across the tuned circuit will be in phase with i_c (since the circuit is resonant at its natural frequency and hence purely resistive). The current i_L through L_1 will lag on i_c by almost 90° and the e.m.f. v_2 induced in the base circuit will lead or lag on i_L by 90° depending upon the sign of the coupling M; this in turn depends simply upon the direction of connection of coil L_2. This induced e.m.f. acts as an input voltage (V_2) and if this is in phase with collector current i_C energy is being fed back in such a manner that the circuit is self exciting.

There is a multiplicity of possible circuit arrangements involving a transistor amplifier and a resonant *LC* circuit in the feedback path and you will encounter some of these in your practical work. In nearly all of these, frequency is controlled by *C*. If the oscillator is going to be used at one single frequency it is usual to employ a fixed, stable capacitor for *C* and adjust *L* by means of a ferrite core, this core being sealed in position after the frequency has been set. In oscillators required to cover a wide range of frequency, the value of *L* may also be made variable by switching in different coils to provide an overlapping set of frequency ranges, each tuned by the single capacitor.

The *LC* tuned circuit type of oscillator can be used to generate frequencies up to many tens of megahertz without difficulty, but for low frequency operation (of the order of a few hundred hertz and less) the values of both *L* and *C* become large, and resistive losses (as well as physical bulk) particularly in the inductor, become prohibitive. It is then preferable to use resistance-capacitance feedback networks, and the oscillator forms which result go under the general name of *RC* or *phase-shift* oscillators. The output waveform from these oscillators is also sinusoidal.

It might seem on first consideration that a network made up of resistors and capacitors would not only be non-resonant (in the sense that *LC* circuits are resonant) but would also introduce a considerable resistive loss. Both of these points are true: a resistor-capacitor network cannot exhibit resonance in the usual meaning of the term and there is bound to be resistive loss. But we must keep in mind that we are concerned with *phase shift* in our feedback network, not resonance; and resistive loss can always be made good in the amplifier part of the loop.

Recalling the work on resistance-capacitance networks in Section 1, we noted there that such networks can change the phase of a signal. This is sometimes a nuisance but it can be put to good effect on other occasions. For example refer to *Figure 6.4* which shows that when a

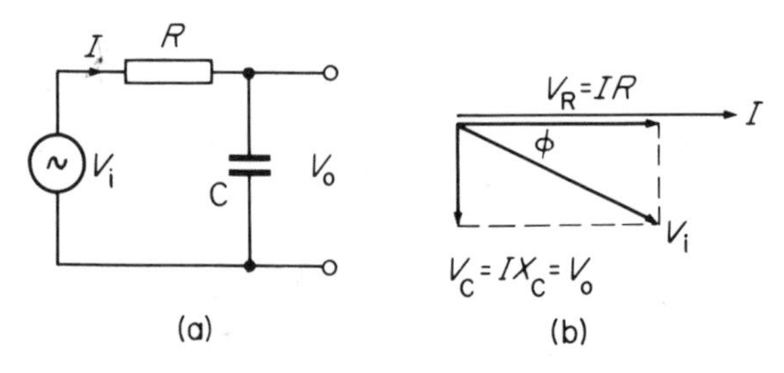

Figure 6.4

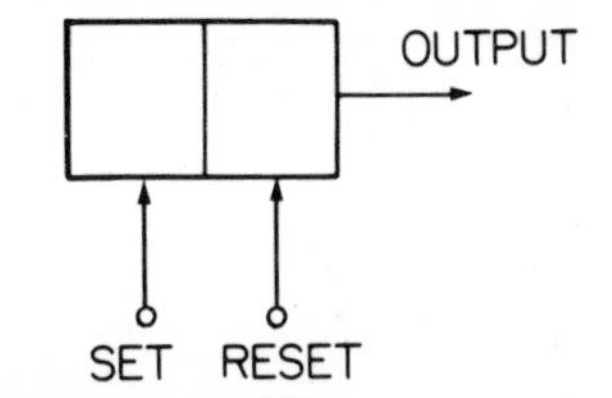

Figure 6.16

The circuit has now changed state and the switch-over has taken place so rapidly that almost immediately after the onset of the input pulse the circuit has completed the action. In this condition the circuit is again perfectly stable *and will remain so* with T_1 switched off and T_2 saturated until a further trigger pulse, appropriately applied, causes the circuit to switch back. This can be achieved by applying a negative pulse to the base of T_2. The input to the base of T_2 is made by way of the RESET terminal as shown in *Figure 6.19.* A convenient way of representing a bistable circuit is shown in *Figure 6.16.*

It would not be easy in practice to interchange the SET and RESET inputs by manual means in order periodically to switch the circuit from one stable state to the other. However, if the bistable is modified to the form shown in *Figure 6.17*, the setting and resetting operation is carried out automatically. Two diodes D_1 and D_2 have been added to the basic circuit and these direct or steer the input pulses to the appropriate transistor base; for this reason they are known as *steering diodes.*

Assume as before that T_1 is switched on and T_2 is cut off. Under these conditions D_2 is *reverse* biased by a voltage almost equal to V_{cc} (check the diagram to make sure you understand how this comes

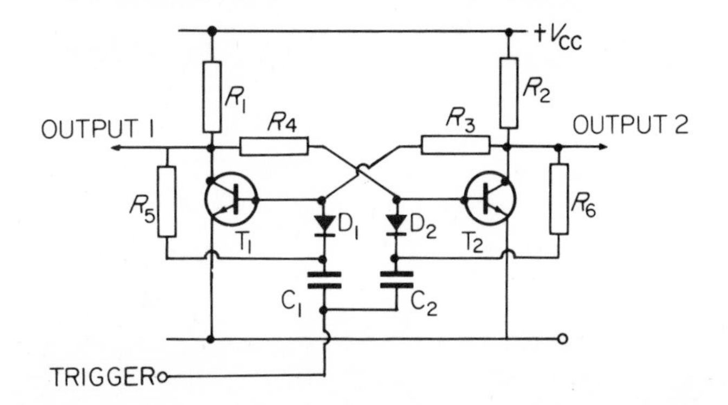

Figure 6.17

about), but D_1 has a small *forward* bias given by the base collector potential of T_1. The input trigger pulses are fed to the diodes through small value capacitors C_1 and C_2 and as a result are differentiated by C_1R_5 and C_2R_6. On the negative edge of the waveform (see *Figure 6.18*), D_1 conducts and puts a bias on to the base of T_1. As T_1 switches off its collector potential rises, T_2 starts to conduct and its collector voltage falls, reducing the base current of T_1. The switch-over action then rapidly follows and ends with T_1 cut off and T_2 fully conducting. The bias condition on the steering diodes is now, of course, also reversed so that the next negative edge of a trigger pulse cuts off T_2 and the circuit reverts to its original state.

Figure 6.18 shows the waveforms of output and input signals for a succession of input pulses. You should particularly notice that the positive spikes produced by the process of differentiation are unable to influence the changeover action of the bistable.

From the waveforms given in the previous figure an important fact can be deduced. This is that an input pulse produces an output signal which remains even when the input has ceased i.e. the circuit 'remembers' that it has received an input signal. So one important application of the bistable circuit is its use as a *memory* element. It

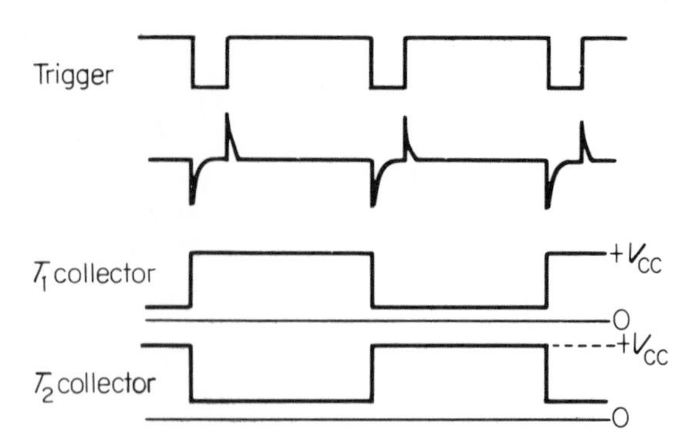

Figure 6.18

has another, perhaps more important application in that it can be used as a binary *counter*. Suppose two bistable circuits are connected in cascade as shown in *Figure 6.19*, the input of the second bistable being derived from the output of the first.

Let a train of negative pulses be applied to terminal A. The output of the first bistable at B will, as *Figure 6.18* previously showed us, be a train of negative pulses of half the frequency of the input train, that is, the circuit has *counted* the input pulses in pairs and *one* output pulse has been obtained for every *two* input pulses. The process can be looked on as division by two. The second bistable will likewise divide its input pulse train by two and hence will give an output pulse at C for every *four* input pulses at A.

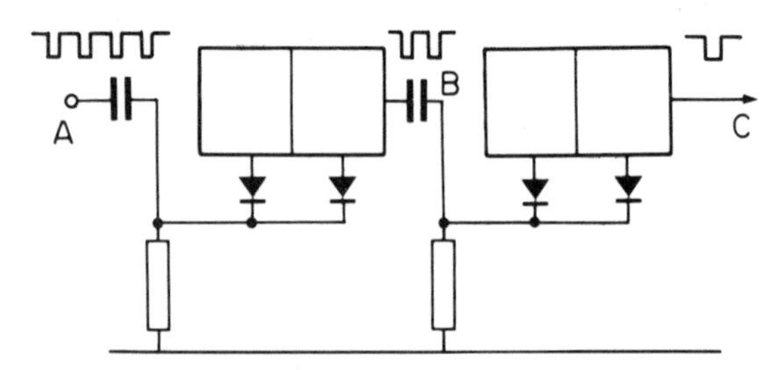

Figure 6.19

By similar reasoning, it is evident that if there were three bistable stages in cascade there would be one output pulse for every eight input pulses—if there were four stages, sixteen input pulses would be necessary to give one output pulse. A series of bistable stages will consequently count input pulses for us in powers of 2, or, remembering your earlier work on logical systems, in *binary notation*.

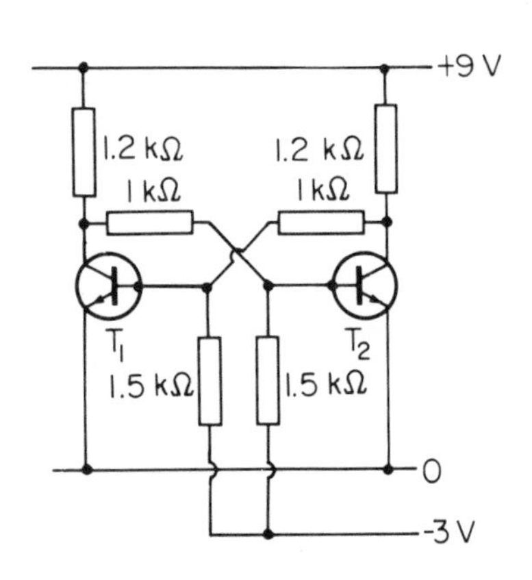

Figure 6.20

Example (12) A bistable multivibrator is shown in *Figure 6.20*. Both transistors are identical and have a current gain of 100. When either transistor is fully on the base current is 50 μA. Ignoring leakage currents, calculate:

(a) The potential between collector and emitter of an 'on' transistor;

(b) The potential between base and emitter of an 'off' transistor;

(c) The potential between base and emitter of an 'on' transistor.

The best way of dealing with this sort of problem is to draw diagrams showing the various current tracks when one transistor is assumed off and the other on. Assuming that T_1 is on, for example, the diagram of *Figure 6.21(a)* can be sketched.

Now base current = 50 μA and current gain (α_E) = 100; therefore the collector current I_C for the 'on' transistor = 100 × 50 μA = 5 mA.

Using Kirchhoff and working in kΩ and mA, we have

$$I_1 + 1.5I_1 + 1.2(5 + I_1) = 12$$

$$\therefore \quad 3.7\ I_1 + 6 = 12$$

$$\therefore \quad I_1 = \frac{6}{3.7} = 1.62 \text{ mA.}$$

(a) Volt drop across the 1.2 kΩ collector load resistor

$$= 1.2(1.62 + 5) = 7.95 \text{ V}$$

$$\therefore \quad V_{CE} \text{ of } T_1 = 9 - 7.95 = 1.05 \text{ V}$$

(b) Volt drop across 1.5 kΩ base resistor

$$= 1.5 \times 1.62 = 2.43 \text{ V}$$

$$V_{BE} \text{ of } T_2 = -3 + 2.43 = -0.57 \text{ V}$$

This bias ensures that T_2 is held 'off'.

Referring now to *Figure 6.21(b)* and applying Kirchhoff again:

$$1.5I_2 + (1 + 1.2)(I_2 + 0.05) = 12$$

$$\therefore \quad 3.7I_2 + 0.11 = 12$$

$$\therefore \quad I_2 = \frac{11.89}{3.7} = 3.21 \text{ mA}$$

(c) Volt drop across 1.5 kΩ base resistor

$$= 1.5 \times 3.31 = 4.81 \text{ V}$$

$$\therefore \quad V_{BE} \text{ of } T_1 = -3 + 4.81 = 1.81 \text{ V}$$

This bias ensures that T_1 is held 'on'.

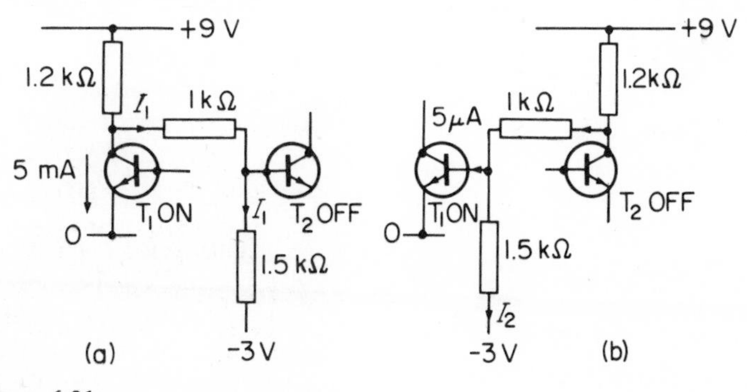

Figure 6.21

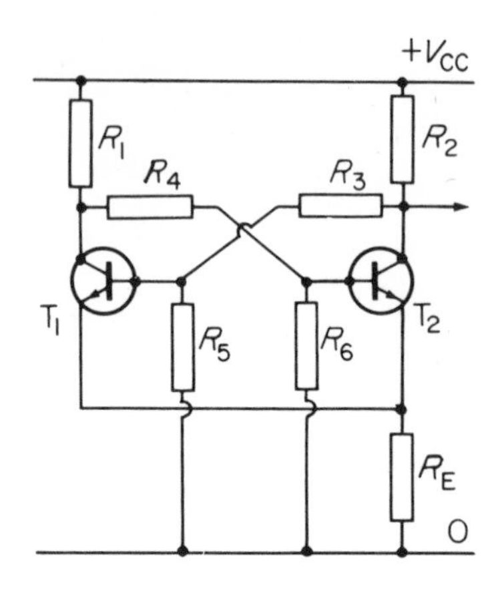

You should notice from the previous example the use of a negative supply rail. This is a more practical method of designing a bistable multivibrator than was the simpler basic circuit already discussed where the base resistors returned to the zero (earth) line. The use of a negative rail ensures that one transistor is always firmly biased off when the other is conducting.

To avoid the use of a negative supply, the emitters are often returned to the zero line through a common resistor R_E as shown in *Figure 6.22*. This puts the emitter potential above earth by the voltage drop across R_E as a result of the current flowing through the 'on' transistor; hence the earthy ends of resistors R_5 and R_6 are effectively negative with respect to the emitters. This circuit is known as the *emitter-coupled* bistable.

PROBLEMS FOR SECTION 6

(13) Complete the following statements:

(a) The amplifier is that part of a feedback loop that determines the . . . . of the generated oscillation.

(b) In a phase advance network, significant phase shift occurs when the frequency is . . . .

(c) In a phase shift oscillator, the ladder network determines the . . . . of the generated waveform.

(d) If the time constant of an astable multivibrator couplings are halved, the output frequency will be . . .

(e) In *Figure 6.10*, when T_1 is conducting, its collector voltage will be at its . . . . value.

(f) In *Figure 6.10*, when T_2 is conducting, the collector voltage of T_1 will be . . . .

(14) Referring to the waveforms of *Figure 6.11*, on what time-constant does the interval between instants (1) and (2) depend?

(15) In *Figure 6.10*, sketch the path by which C_1 charges during the interval during which T_1 is switched off.

(16) If in the circuit of *Figure 6.10* $C_1 = C_2 = 5000$ pf, $R_1 = R_2$ 50 kΩ, what will be the generated square wave frequency?

(17) Are the following statements true or false?

(a) A relaxation oscillator always generates rectangular waves.

(b) An astable multivibrator has two quasi-stable states.

(c) It is not possible to design an oscillator with a common-base amplifier because such an amplifier has zero phase shift.

(18) In the circuit of *Figure 6.10*, calculate the collector-emitter voltage for T_2, the 'off' transistor. (You can refer to worked example (12) if you wish).

(19) Now deduce the maximum and minimum amplitude of the output wave generated by the bistable of *Example (18)*.

(20) Refer to *Figure 6.19*. If pulses arrive at terminal A with a frequency of 5 kHz, what will be the frequency of the output pulses at C?

(21) How many binary (bistable) stages would you require to give a count of pulses between 0 and 128?

(22) For the monostable circuit of *Figure 6.13*, draw the portion of the circuit that affects the switch-on and switch-off of T_1 and describe the cause of the switching.

(23) Confirm that the frequency of a generated square wave in an astable multivibrator is given by $f = 1/1.38CR$, where $C = C_1 = C_2$ and $R = R_1 = R_2$.

7 Large signal amplifiers

Aims: At the end of this Unit section you should be able to:
Understand the operation of a Class-A single-ended and push-pull large signal amplifier.
Apply graphical analysis to transformer-coupled Class-A amplifiers for the estimation of power output, efficiency and collector dissipation.
Understand the operation of a Class-B push-pull large signal amplifier.
Apply graphical analysis to transformer-coupled Class-B amplifiers.
Sketch and explain the principles of complementary and transformer-less amplifiers.
Explain the reasons for distortion arising in power amplifiers.
State methods of eliminating parasitic oscillation.

A large-signal amplifier, commonly referred to as a power amplifier, is a converter of d.c. power drawn from a power supply source into a.c. power delivered to a load. As these amplifiers are designed to supply very high power to their output loads, rated in watts (or kilowatts in some applications) the main consideration must be the elimination of power wastage in the amplifier, which means that the efficiency and not the power gain is the parameter of major importance.

There are of course maximum theoretical efficiencies applicable to the various classes of operation. In Class A, for example, this figure is 50%, which means that a 100 W amplifier operating in this class will waste at least 50 W in the form of heat within the amplifier. Since the actual figure for efficiency obtained in practice will be less than 50%, Class A is not a particularly attractive mode of operation—at least, not where powers greater than a few watts are concerned.

Class B offers a better proposition; here we shall see that the maximum efficiency is about 78%, which means that only 22 W is wasted as heat out of a total of 100 W.

Class C operation, which is reserved for particular applications and cannot be used in the kind of amplifiers we are going to discuss, can have an efficiency of the order of 85%.

In all classes of operation, all the wasted power has to be dissipated by the output device, transistor or valve, and this sets a limit to the useful power output that a particular device may supply.

When an amplifier is delivering appreciable power to a load, a large amplitude input signal is required to produce the necessary large swings of output voltage and output current which dissipate the power in the load. This means that we are no longer working the output device over a limited portion of its output characteristic curves as we were for small-signal amplifiers. For this reason, power amplifiers are best studied by graphical analysis and basic Ohm's law algebra, the small-signal equations and parameters being no longer valid.

In this section we shall deal only with transistor power amplifiers operating over audio-frequencies. but the information can apply equally well to amplifiers using thermionic valves.

CLASS A AMPLIFIERS

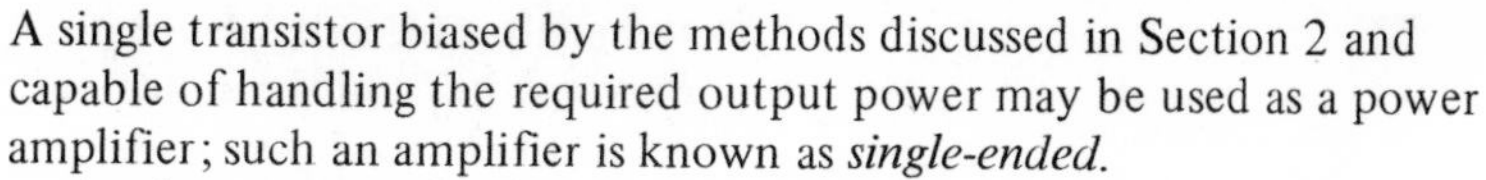

A single transistor biased by the methods discussed in Section 2 and capable of handling the required output power may be used as a power amplifier; such an amplifier is known as *single-ended.*

As the power to be delivered to the load should be as great as possible, it is necessary to ensure that the load is properly matched to the output resistance of the transistor, since the maximum power is transferred from a generator (the transistor) to the load when the proper relationship is established between their resistances. The particular value of load resistance which suits the transistor being used is known as the *optimum* load.

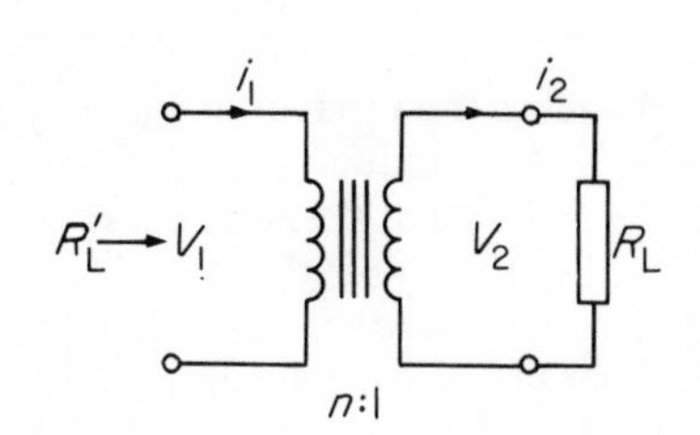

Figure 7.1

It is usual to employ transformer coupling in single-ended amplifier stages, the turns ratio being chosen so that the resistance presented at the primary terminals of the transformer is that required to suit the output transistor. In the transformer of *Figure 7.1* the effective resistance seen at the primary terminals is R_L' where

$$R_L' = \frac{V_1}{i_1}$$

Assuming that the transformer is loss free, then

$$\frac{V_1}{V_2} = n \text{ and } \frac{i_1}{i_2} = \frac{1}{n}$$

where n is the transformation ratio given by T_1/T_2.

$$\therefore \quad R_L' = \frac{n.V_2}{i_2/n} = n^2 . \frac{V_2}{i_2}$$

but

$$\frac{V_2}{i_2} = R_L \text{ the load resistance}$$

$$\therefore \quad R_L' = n^2.R_L \tag{7.1}$$

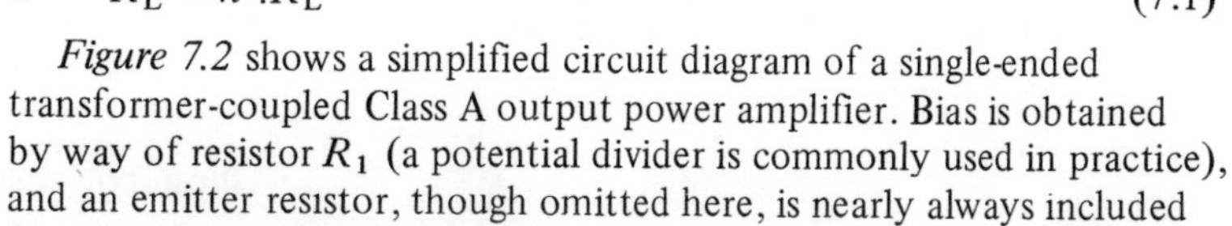

Figure 7.2 shows a simplified circuit diagram of a single-ended transformer-coupled Class A output power amplifier. Bias is obtained by way of resistor R_1 (a potential divider is commonly used in practice), and an emitter resistor, though omitted here, is nearly always included for reasons of thermal stability.

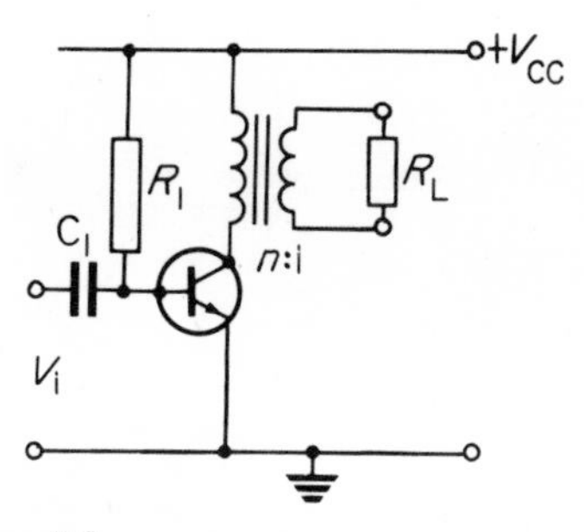

Figure 7.2

Assuming that the d.c. resistance of the transformer primary winding is negligible and that there is no signal input to the base, then the d.c. power supplied at any instant is

$$P_{dc} = V_{cc}.I_Q$$

where I_Q is the steady or quiescent collector current flowing at the selected bias point.

But the power dissipated at the collector at any instant is the product of the collector voltage and this steady current, so

$$P_C = V_Q.I_Q$$

and this is the same as the d.c. power supplied since the collector quiescent voltage, V_Q, is equal to V_{cc}, there being no voltage drop in the transformer winding. Hence the *whole* of the power supplied is dissipated as heat at the collector.

Suppose a sinusoidal signal voltage V_i to be applied to the input

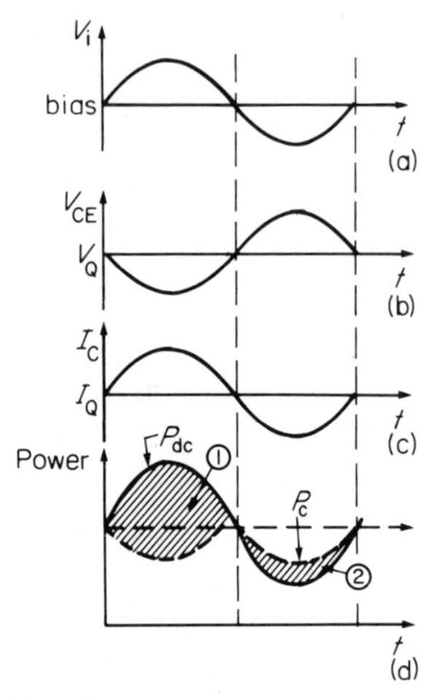

Figure 7.3

terminals. This voltage will vary about the mean base bias as shown in *Figure 7.3(a)* and the collector current I_C will vary in phase with V_i about the quiescent value I_Q (*Figure 7.3(c)*). To this alternating current the transformer primary resistance is not zero (as it is for d.c.) but is given by $n^2.R_L$. Hence the collector voltage V_C will vary in antiphase to the collector current about the quiescent value V_Q ($=V_{cc}$), as shown in diagram (b).

Now the power supplied by the d.c. source is the product of V_{cc} and I_Q, hence the curve of d.c. power supplied is the product of curve (*c*) and the constant V_{cc}. This resultant curve is shown in the full line wave of *Figure 7.3(d)*. The power dissipated at the collector is the product $I_Q V_Q$, hence the curve of power dissipated is the product of curves (b) and (c). Since these are phase-opposed, their product will be wholly negative, and the resultant wave this time is shown in broken line in diagram (*d*).

The area under a power-time graph represents energy, so by shading in the areas between the two curves we obtain the energy representations of what happens during each half-cycle of input signal between the power supplied (P_{dc}) and the power dissipated at the collector (P_C). As the areas are not equal, some of the energy supplied must go elsewhere other than as dissipation, and this must obviously be as energy delivered to the load.

During the positive half-cycle of input signal, energy given to the load is positive i.e. energy given by the source is *greater* than the collector dissipation. During the negative half-cycle the energy given by the source is *less* than the collector dissipation. The useful power output into the load is thus represented by area 1 minus area 2, that is

$$\text{Power in the load } P_L = P_{dc} - P_c$$

Since P_{dc} is constant, *the collector dissipation is greatest when the input signal is zero* and $P_L = 0$. All transistors (and valves) have a figure quoted for the maximum safe power they can dissipate and care must be taken to ensure that P_C never exceeds this.

Now

$$\text{Efficiency} = \frac{\text{mean power supplied to load}}{\text{mean power supplied by source}}$$

$$= \frac{\text{power in the load}}{\text{power in the load + total dissipation}}$$

$$= \frac{P_L}{P_L + P_C}$$

Strictly, the total dissipation must include the small resistive loss occurring in the transformer primary, but this is usually small and we will normally assume that the collector dissipation represents the only power wastage.

Example (1) A Class A power amplifier draws a quiescent collector current of 0.5 A from a 20 V supply and delivers a signal power of 3.5 W to the collector load. Calculate: (a) the efficiency; (b) the collector dissipation.

$$P_{dc} = V_{cc}.I_Q = 20 \times 0.5 = 10 \text{ W}$$

(a) $$= \frac{P_L}{P_{dc}} = \frac{3.5}{10} = 0.35\ (35\%)$$

(b) $$P_C = P_{dc} - P_L$$
$$= 10 - 3.5 = 6.5 \text{ W.}$$

(2) A Class A amplifier has an efficiency of 25% and is required to deliver 5 W of signal power to the load. What is the collector dissipation and the d.c. supply power?

Graphical analysis of Class A

The graphical determination of the power output and efficiency of a large-signal amplifier is carried out in exactly the same way as that used for finding the voltage or current gains of small-signal amplifiers, that is, by using the signal (dynamic) load line drawn on the output characteristics of the transistor. In the case of small-signal voltage amplifiers, having a resistance as the collector load, the a.c. and d.c. load lines were very often almost coincident, but in the case of power amplifiers, they differ considerably.

If, as we have assumed, the d.c. resistance of the transformer primary winding is negligible, the effective d.c. load on the transistor is zero. So the d.c. load line is vertical, starting from the point $V_Q = V_{cc}$ as shown in *Figure 7.4*. At the signal frequency the effective load resistance $R_L{}'$ is $n^2.R_L$, hence the a.c. load line will have a gradient of $-1/R_L{}'$ and will intersect the d.c. load line at the operating point P.

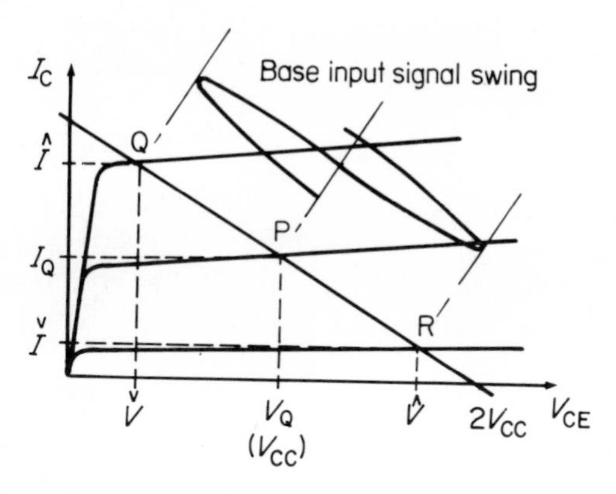

Figure 7.4

When an alternating signal is applied to the base, the base current varies about its steady bias value and the corresponding excursions of output voltage and current can then be determine by projecting from the intersection of the a.c. load line and the limiting base current curves, points Q and R in the diagram, to the co-ordinate axes. You should notice that the collector voltage of a transformer-coupled stage can swing *above* the d.c. supply voltage V_{cc}. In fact, the load line crosses the V_{CE} axis at a value $2V_{cc}$.

Let $\hat{V}$, V be the maximum and minimum excursions of the collector voltage swing about the mean V_Q, and let $\hat{I}, \check{I}$ be the maximum and minimum values of the corresponding collector current swing about the mean I_Q. Then the peak value of the voltage wave is $(\hat{V} - V)/2$ and the r.m.s. value is thus divided by 2, so

$$V_{rms} = \frac{\hat{V} - \check{V}}{2\sqrt{2}}$$

Similarly, the r.m.s. value of the current wave is

$$I_{rms} = \frac{\hat{I} - \check{I}}{2\sqrt{2}}$$

The power output is the product of the r.m.s. voltage and current, hence

$$P_L = \frac{(\hat{V} - \check{V})(\hat{I} - \check{I})}{2\sqrt{2} \quad 2\sqrt{2}} = \frac{(\hat{V} - \check{V})(\hat{I} - \check{I})}{8} \tag{7.2}$$

Let us assume that the whole of the load line can be utilised so that the greatest possible output power is obtainable. In such a case, as *Figure 7.4* shows, $\check{V}$ and $\check{I}$ would be zero, $\hat{V}$ would be $2V_{cc}$ and $\hat{I}$ would be $2I_Q$. Equation (7.2) above then reduces to

$$P_{L(max)} = \frac{2V_{cc}.2I_Q}{8}$$

But the d.c. power supplied = $V_{cc}.\hat{I}_Q$

$$\therefore \quad \eta_{max} = \frac{P_{L(max)}}{P_{dc}} = \frac{2V_{cc}.2I_Q}{8V_{cc}.I_Q} = \frac{1}{2} \text{ or } 50\%$$

This is the theoretical maximum efficiency for a transformer-coupled Class A amplifier. In practice this figure can never be reached since point Q and R cannot be made to coincide with the extremes of the load line; however, efficiencies of 45% are possible with transistor stages, though with valves, figures of 20% to 30% are more likely.

(3) Power can be expressed as I^2R or as V^2/R. Using this hint, express the power in the load of a Class A amplifier in terms of

(a) $\hat{I}, I_Q$ and R_L' (b) $\hat{V}, \check{V}$ and R_L'

(4) If a Class A amplifier has a *resistance* as its collector load (that is, the load is not transformer-coupled), show that the efficiency cannot be greater than 25%. (*Hint: What is the greatest collector voltage swing in this case?*)

THE MAXIMUM DISSIPATION CURVE

As we have already noted, transistor manufacturers quote a maximum collector dissipation figure for their products. It is necessary to ensure that this figure is not exceeded continuously when the transistor is in operation. A maximum power dissipation curve can be superimposed on the output characteristics and will have the equation

$$V_C.I_C = P_{C(max)}$$

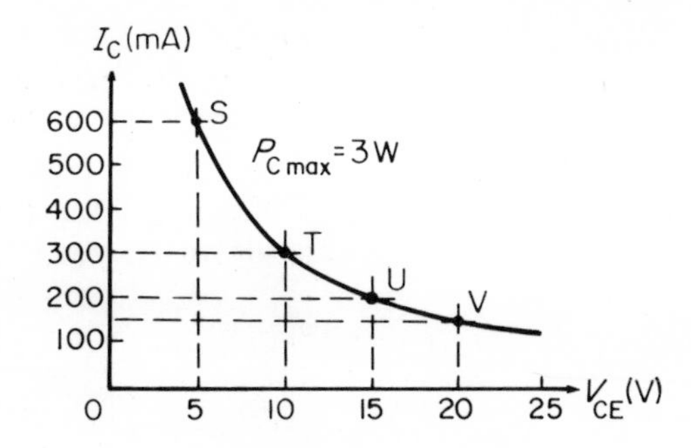

Figure 7.5

Figure 7.5 shows how the curve is plotted. Suppose $P_{C\ max}$ is stated to be 3 W. Then on the output characteristic axes a number of points may be plotted where the product $V_C.I_C$ = 3 W. For example, at V_C = 5 V, the collector current I_C for a dissipation of 3 W will be 3/5 = 0.6 A; the point V_C = 5 V, I_C = 0.6 A therefore lies on the dissipation curve and is shown as point S in the diagram. The remaining points T, U and V, corresponding to V_C = 10 V, I_C = 0.3 A, V_C = 15 V, I_C = 0.2 A and V_C = 20 V, I_C = 0.15 A respectively can be located similarly. Four such points are usually sufficient. A *smooth* curve joining these points then defines the maximum dissipation curve corresponding, in this example, to 3 W. The curve is a hyperbola.

Continuous operation of the transistor outside the boundary of this curve is not permissible. The load line does not have to touch the hyperbola but it is clear that the available power output is increased as it approaches it as a tangent at the operating point. Ideally, the condition shown in *Figure 7.6* should be aimed for, the load line being

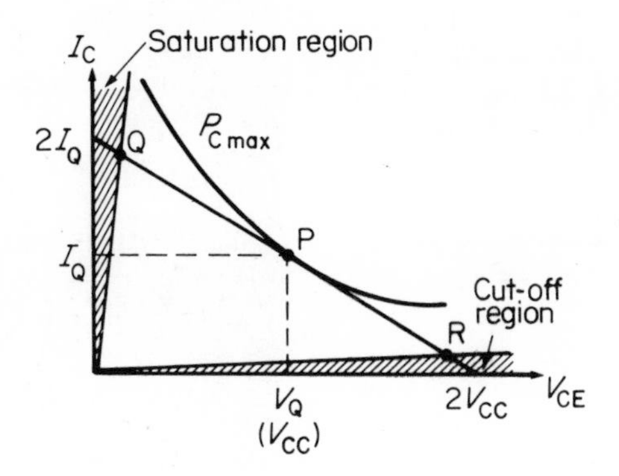

Figure 7.6

tangential to the hyperbola at the operating point P. The shaded areas represent the excursion limits of the signal if distortion is to be avoided; beyond point Q on the load line saturation sets in, beyond point R at the other end the transistor cuts off. Any signal excursions into these regions lead to 'clipping' and hence distortion.

Example (5) A transistor having $P_{C\ max}$ = 0.5 W has V_{cc} = 10 V is transformer-coupled to a 3 Ω loudspeaker load. Assuming 50% efficiency calculate: (a) the output power in the load; (b) the required output transformer ratio; (c) the least primary inductance required if the 3 dB low-frequency response point is to be 100 Hz.

(a) At 50% efficiency, the load power will be one half of the transistor dissipation. Hence $P_L = 0.25$ W.
(b) We require to find R_L' in terms of V_{cc} and the load power, as we are not given the transformation ratio.

$$\text{Now } I_Q = \frac{V_{cc}}{R_L'} \quad \text{and} \quad P_{dc} = V_{cc} \cdot I_Q$$

Keep in mind that we are only using Ohm's law in these problems. Combining these equations we get

$$R_L' = \frac{V_{cc}^2}{P_{dc}} = \frac{V_{cc}^2}{2P_L}$$

since $P_L = P_{dc}/2$ at 50% efficiency. Inserting the given figures

$$R_L' = \frac{10^2}{2 \times 0.25} = 200\ \Omega$$

Hence the required turns-ratio is found from $R_L' = n^2 R_L$

$$n = \sqrt{\frac{200}{3}} = \sqrt{66.7} = 8.2$$

The transformer is clearly 8.2 : 1, step-down.

(c) The 3 dB response point occurs when reactance = resistance, or when the phase angle = 45°, remember?

$$\therefore \quad \omega L = R_L' \text{ or } L = \frac{R_L'}{\omega}$$

$$\therefore \quad L = \frac{200}{2\pi \times 100} = \frac{1}{\pi} = 0.318 \text{ H}$$

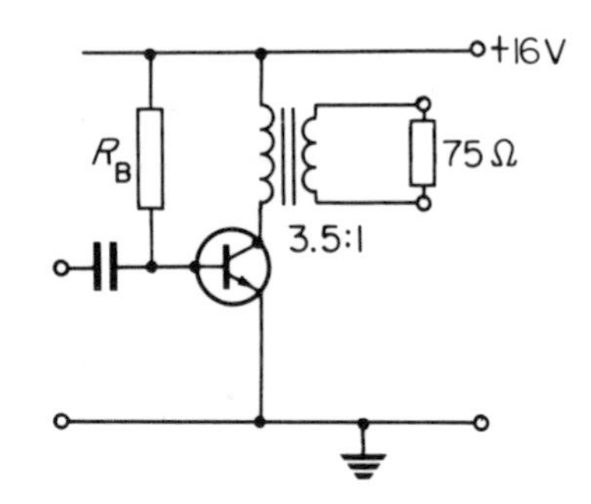

Figure 7.7

Example (6) The output characteristics of a transistor having a maximum collector dissipation of 0.25 W used in the circuit of *Figure 7.7* are given below:

V_{ce} (V)	2	20	for I_B (mA)
I_C (mA)	0.2	1.0	0
	5.5	6.5	0.1
	17.0	19.0	0.3
	29.5	32.5	0.5

Plot the characteristics and superimpose on them (i) the maximum dissipation hyperbola; (ii) the d.c. load line; (iii) the a.c. load line. Assuming a sinusoidal input signal of 0.25 mA peak current, select a suitable operating point and hence estimate: (a) the power developed in the load resistor; (b) the collector efficiency; (c) the value of bias resistor R_B.

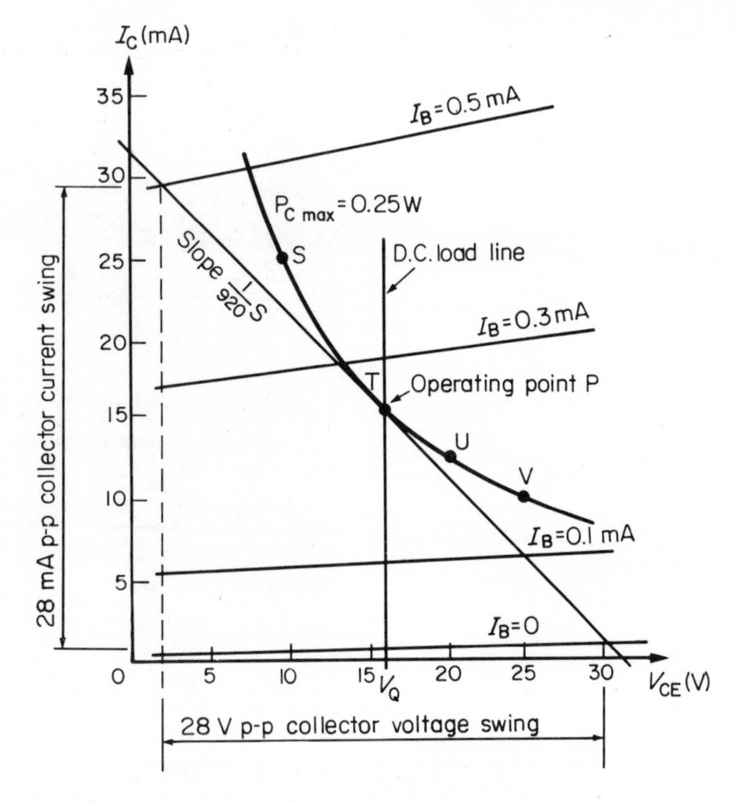

Figure 7.8

The characteristics are plotted in *Figure 7.8*. Since $P_{C\ (max)}$ = 0.25 W, points on the hyperbola can be plotted conveniently for values of V_C equal to 10, 15, 20 and 25 V, leading to corresponding values of I_C of 25, 16.7, 12.5 and 10 mA respectively. These points are indicated as S, T, U and V on the diagram and joined with a smooth curve.

The d.c. load line will be vertical, passing through the point on the horizontal axis where V_C = 16 V (= V_{cc} of course).

The a.c. load line will have a gradient of $-1/R_L'$ where $R_L' = n^2 R_L = 3.5^2 \times 75 = 920\ \Omega$. The gradient is therefore $-1/920$ S. This line should ideally touch the dissipation curve at the operating point where V_C = 16 V; hence this is drawn on the diagram with the required gradient and passing through the d.c. load line also at the operating point. The corresponding quiescent collector current is seen to be about 15 mA with the base biased at approximately 0.25 mA. Notice that we were not given an output characteristic for I_B = 0.25 mA and this figure is an estimation for the position of the operating point between the I_B = 0.1 mA and I_B = 0.3 mA characteristics.

With the 0.25 mA input peak signal superimposed on the base bias of 0.25 mA, the base will swing from I_B = 0 to I_B = 0.5 mA.

The remainder of the problem can now be completed from a study of the graph.

(a) We will calculate the power output from the collector voltage swing and the load resistance. From the graph the peak-to-peak collector voltage swing is about 28 V. Hence the r.m.s. value is $28/2\sqrt{2} \simeq 10$ V.

The load power

$$P_L \simeq (10)^2 \cdot \frac{1}{920} \text{ W}$$

$$= 0.108 \text{ W}$$

(b) $$\frac{\text{power delivered to load}}{\text{d.c. power supplied}} = \frac{P_L}{V_{cc} \cdot I_Q}$$

$$= \frac{0.108}{16 \times 15 \times 10^{-3}} \times 100\% = 45\%$$

(c) The operating point is located at $I_B \simeq 0.25$ mA.

$$\therefore \quad R_B \simeq \frac{V_{cc}}{I_B} \simeq \frac{16}{0.25} \times 10^3 \simeq 64 \text{ k}\Omega$$

CLASS A PUSH-PULL AMPLIFIERS

By using two transistors in what is known as a push-pull amplifier, a greater power output than is possible from a single-ended stage becomes possible. By connecting two transistors in Class A push-pull as shown in *Figure 7.9* and driving them with input signals that are equal in magnitude but opposite in phase, the outputs can be combined in a single transformer to give an increased power in the load.

With no signal input the normal quiescent collector current flows in each transistor, its value determined by the base bias. From *Figure 7.10(a)* it is clear that the two currents flow in the two sections of the transformer primary in opposite directions.

Providing that the transformer is accurately centre-tapped and the transistors have identical collector currents, there will be no resultant

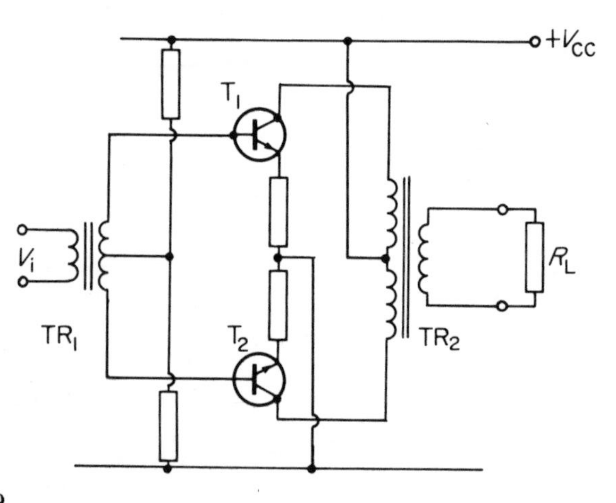

Figure 7.9

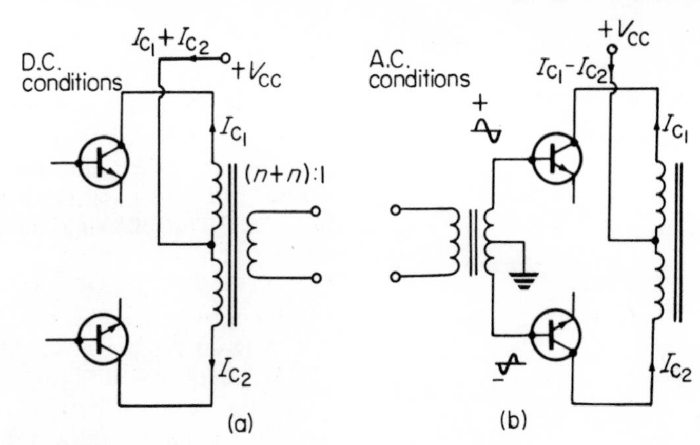

Figure 7.10

d.c. magnetisationof the core. This means that a smaller transformer can be used and a high primary inductance maintained. The inclusion of small value emitter resistors as shown in *Figure 7.9* helps the d.c. balance of the circuit.

When a signal is developed across the secondary winding of the input or *phase-splitter* transformer TR_1, the transistor base inputs are anti-phase to each other. Thus, as *Figure 7.10(b)* shows, if T_1 base goes positive, its collector current will increase while at the same time the base of T_2 goes negative and its collector current decreases by a similar amount. Hence the name 'push-pull'. It is seen that these a.c. current changes, unlike the d.c. condition, are *additive* in the transformer primary but that they cancel out in the V_{cc} supply line. Hence no signal voltages get into the supply line and the need for heavy decoupling is reduced. Further, the identical even harmonics, if present, will cancel out as well.

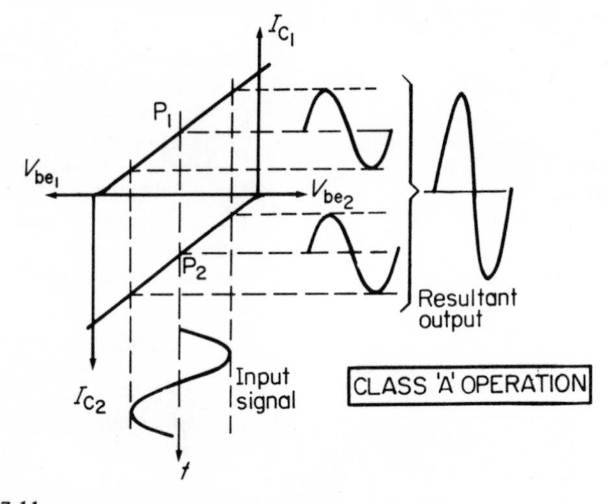

Figure 7.11

Figure 7.11 illustrates the action of Class A push-pull with respect to the dynamic transfer characteristics of the transistors. P_1 and P_2 are the operating points for transistors T_1 and T_2 respectively. The individual outputs are then summed in the transformer primary winding.

Matching Care has to be taken with push-pull circuits to get the transformer matching right. In Class A operation the load current is doubled over that of the single-ended stage but the voltage is unchanged. If each transistor is to look into the same load as it did in the single-ended case, the value of R_L must be halved, assuming that we effectively use two primaries in series, that is, an $n : 1$ transformer becomes an $(n + n) : 1$. The following example will illustrate this point for you.

Example (7) A certain transistor has a collector dissipation of 2 W. A 20 V supply is used and the transistor is coupled to a 8 Ω loudspeaker load through an ideal transformer. Calculate the output power and the transformer ratio for (a) a single-ended Class A amplifier; (b) a push-pull Class A amplifier.

(a) The greatest output we can obtain from the single-ended stage is one-half the dissipation i.e. 1 W. To obtain this, the whole of the load line must be utilised so that the peak collector voltage swing will equal the supply voltage V_{cc}. From the solution of *Example (5)* above

$$R_L' = \frac{V_{cc}^2}{2P_L} = \frac{20^2}{2} = 200\ \Omega$$

Hence

$$n = \sqrt{\frac{R_L'}{R_L}} = \sqrt{\frac{200}{8}} = 5:1$$

(b) In the push-pull connection the two transistors will dissipate 4 W and provide an output of 2 W. The peak collector voltage remains at 20 V, so the load seen across *either* half of the primary is now 400/4 = 100 Ω. Hence the ratio of each half primary to secondary is

$$\sqrt{\frac{100}{8}} \simeq 3.5 : 1$$

or a ratio of 7 : 1 across the *entire* primary to secondary winding.

This last part of the problem could be looked at by considering the two transistors to be in series (as seen from the primary) so that they provide twice the voltage but the same current. The required load right across the primary is then 400 Ω i.e. double the number of turns but four times the resistance. The ratio now is

$$\sqrt{\frac{400}{8}} \simeq 7 : 1$$

the same solution as before.

To summarise: besides the increased power output, other advantages arise from Class A push-pull operation:

(i) As the d.c. collector currents flow in opposite directions in the output transformer primary, no magnetisation of the core occurs and the inductance remains high.

(ii) The signal currents add in the transformer primary, but the resultant a.c. current flowing from the centre-tap through the supply line is zero. Therefore feedback through the supply line to the earlier stages is eliminated.

(iii) Similarly, all second and higher order even harmonics are cancelled out.

The efficiency of a Class A push-pull amplifier, like a single-ended stage, has a theoretical maximum of 50%. If a higher efficiency is required, Class B operation is necessary.

CLASS B AMPLIFIERS

Figure 7.9 can equally well be used to illustrate a push-pull circuit arrangement for Class B operation. Centre-tapped input and output transformers are used as before but the difference this time is that the two transistors are biased to cut-off. The operating point of each is consequently positioned at the extreme lower end of the dynamic load line as *Figure 7.12* illustrates and not at the centre as it was for Class A.

The *whole* of the load line now becomes available for the signal voltage and current excursions. Under this condition $I_C = 0$ and $V_Q = V_{cc}$. The load line then joins the point *P* with $I_C = I_{C(max)}$, $V_{CE} = 0$. Then $R_L' = V_{cc}/I_{C(max)}$.

An input sinusoidal signal now simultaneously drives one transistor on and the other further into cut-off, so that the transistors conduct on alternate half-cycles to give consecutive half-wave current pulses (just like the output of a half-wave rectifier) flowing in opposite directions in the primary winding of the output transformer. The transformer combines these half-wave inputs to give a full wave output across the load resistance, see *Figure 7.13*.

Although each transistor provides half the output power, the load presented to the transistor which happens to be conducting must be such that the *total* power is dissipated over its half cycle.

Figure 7.14 illustrates the action of Class B push-pull with respect

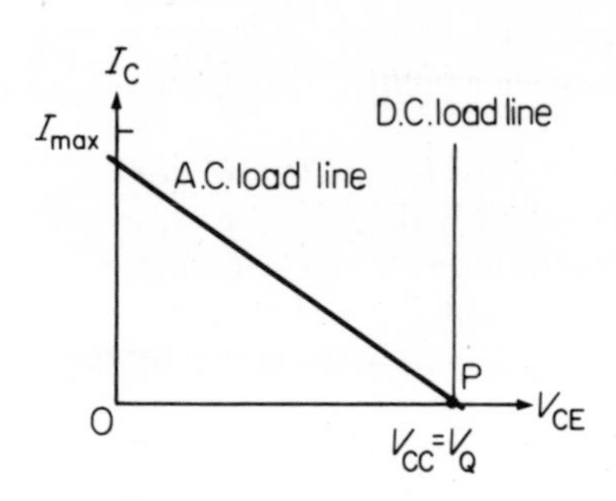

Figure 7.12

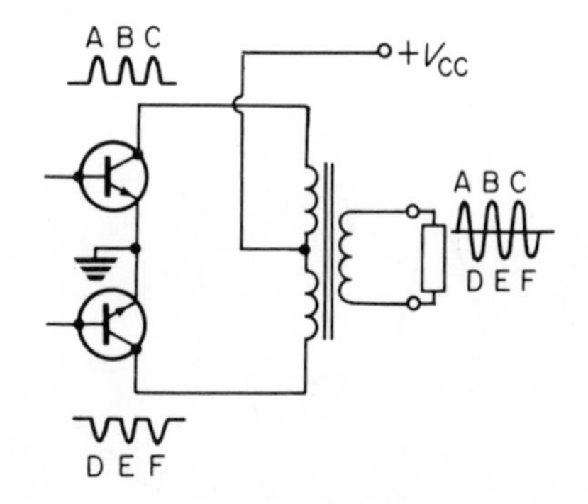

Figure 7.13

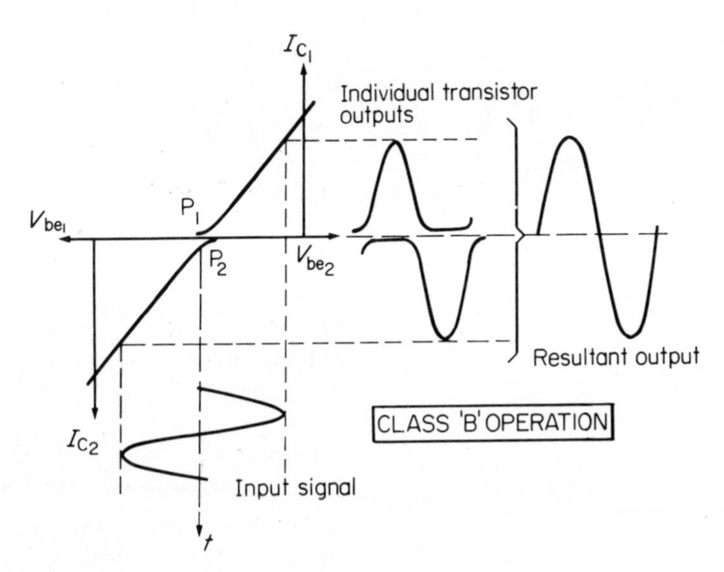

Figure 7.14

to the dynamic transfer characteristics of the transistors. P_1 and P_2 are the operating points for transistors T_1 and T_2 respectively. The individual half-wave outputs are then summed in the transformer primary winding. You should compare this diagram for Class B with that for Class A shown in *Figure 7.11.* Make a point of noticing, by the way, that the operating points in *Figure 7.14* are *not quite* at collector current cut-off but are positioned so that the dynamic transfer characteristics form together a single straight line. We will return to this point a little later on.

Efficiency of Class B

A higher output power and efficiency are obtainable with Class B operation because, in the absence of a signal, the transistors being biased to cut-off, draw negligible current and so there is negligible quiescent power dissipation. Of course with this class of operation there is no question of using a single-ended amplifier, the output would be grossly distorted.

To find the maximum efficiency of a Class B amplifier we need consider only one of the transistors working with a half sine wave output. Both the load current and load voltage are half sine waves, the supply current is also a half sine wave, but V_{cc} is of course constant. We can represent the power conditions in the circuit by the waveforms shown in *Figure 7.15.*

For the d.c. supply power, this is simply the product of the supply voltage V_{cc} and the *mean* value of the load current; during each half cycle this current is $\hat{I}/\pi$. Hence the power from the d.c. supply, as *Figure 7.15(a)* illustrates, is

$$P_{dc} = V_{cc} \cdot \frac{\hat{I}}{\pi}$$

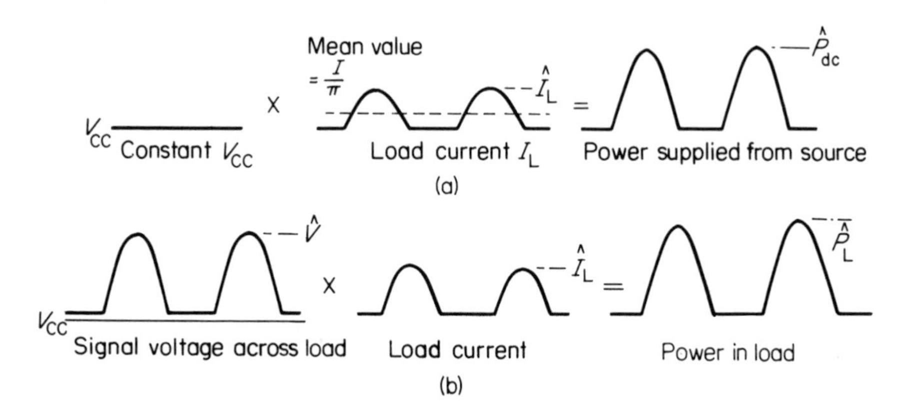

Figure 7.15

For the signal power in the load, we can derive this as the product of the load voltage V_L and the load current I_L, and this is illustrated in *Figure 7.15(b).* Then the load power delivered by a stage that is driven through its total range (cut-off to saturation) is given by

$$P_L = \frac{1}{2}\left[\frac{\hat{V}}{\sqrt{2}} \cdot \frac{\hat{I}}{\sqrt{2}}\right] = \frac{\hat{V}.\hat{I}}{4}$$

Notice that the half factor appears because we are dealing with a half sine wave. Hence the maximum efficiency for each transistor and therefore for the amplifier as a whole is

$$\frac{P_L}{P_{dc}} = \frac{\pi \hat{V}}{4V_{cc}}$$

From *Figure 7.16*, if the whole of the load line is utilised, $\hat{V}$ would equal V_{cc}, and so

$$\eta = \frac{\pi V_{cc}}{4V_{cc}} = \frac{\pi}{4} = 0.785 \text{ (or 78.5\%)}$$

This is the maximum theoretical value for η. In practice the load line cannot be fully utilised, particularly at the saturation end as shown in *Figure 7.16*, so the efficiency is always less than 78.5%. Figures of up to 70% are quite feasible with transistors.

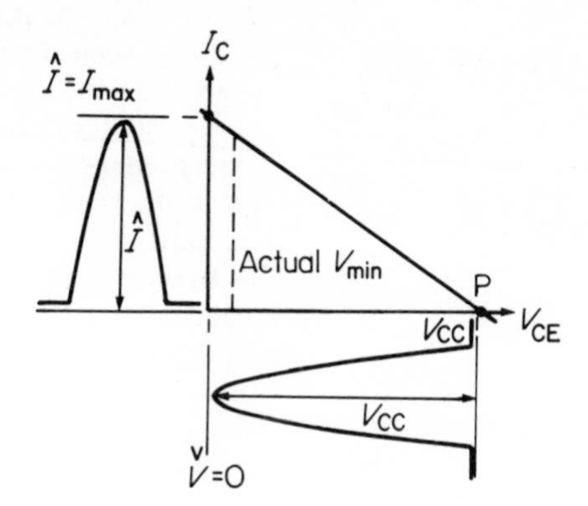

Figure 7.16

Clearly the collector dissipation increases as the signal input increases. We can express the dissipation directly in terms of P_L and η, for

$$P_C = P_{dc} - P_L$$

$$= P_L \left[\frac{P_{dc}}{P_L} - 1\right]$$

$$= P_L \left[\frac{1}{\eta} - 1\right]$$

so for $\eta = 0.785$,

$$P_C = 0.27\, P_L$$

It can be proved that the ratio

$$\frac{\text{maximum output}}{\text{maximum dissipation}} \simeq 2.5$$

for a Class B stage. You will recall that for Class A this ratio was only 0.5. Therefore Class B operation is a much better proposition all round.

(8) Assuming that a Class B amplifier is working at full efficiency, find (a) the peak load power (b) the peak d.c. supply power , in terms of V_{cc} and R_L'.

(9) Repeat problem (7) for Class B operation. (*Hint: keep the ratio mentioned above in mind here*).

(10) Complete the following sentences:
(a) With no signal input, both P_{dc} and P_c are . . . .
(b) In Class B operation, the output transformer effectively feeds the . . . . of the transistor output signals to the load.

TRANSFORMERLESS CIRCUITS

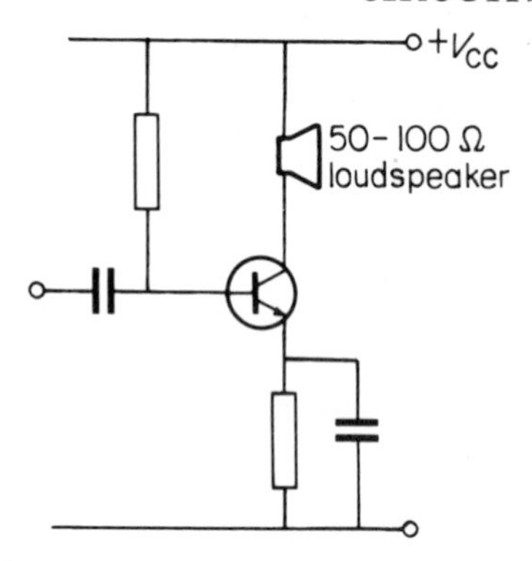

Figure 7.17

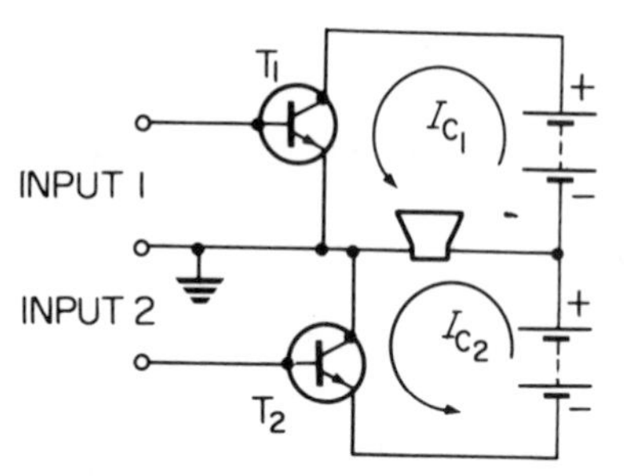

Figure 7.18

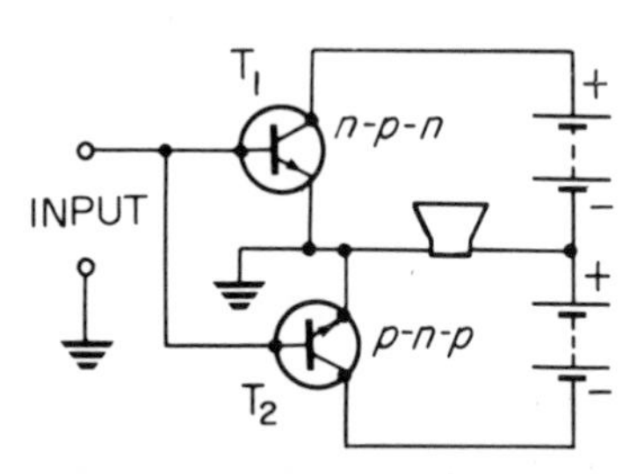

Figure 7.19

Most audio-frequency amplifiers are designed to drive a loudspeaker output; such speakers usually have an impedance between 3 and 50 Ω, which can be considered to be resistive. So far we have considered such loads to be transformer-coupled to the output amplifier, and although such coupling is necessary with thermionic valves, it is not so with transistors.

The ability to do away with the output transformer and couple the loudspeaker directly to the output transistors has led to the development of a variety of circuits. These have been given names such 'transformerless' or 'complementary' output stages.

In simple amplifiers, a Class A amplifier can have the loudspeaker load connected directly into the collector circuit as shown in *Figure 7.17*. Such speakers usually have a resistance of 50 to 100 Ω and a transformer is unnecessary because the a.c. resistance of the load is sufficiently high to produce a reasonable match.

For push-pull amplifiers, the simplified circuit of *Figure 7.18* provides a means of doing away with the output transformer, but antiphase input signals are still required on the bases of the transistors. The circuit is essentially a bridge with the loudspeaker load connected as the centre arm.

The transistors shown are both *n-p-n* types (but *p-n-p* types may be used with reversed supply polarities), and assuming that they are biased to cut-off and so operating in Class B, they switch on and off alternately on each input cycle. The resulting collector currents, indicated by I_{C1} and I_{C2} consequently act in opposite directions in the speaker load (as they did in the transformer primary in the earlier discussion) and reconstitute the complete output wave across the load resistance.

(11) If each transistor is biased to operate in Class A, explain how the circuit will now reproduce the complete signal wave across the load.

(12) With the emitters earthed as shown in *Figure 7.18*, the transistors are in common-emitter configuration. What would be the mode if the centre point of the supply batteries was earthed instead?

A further step is now possible that will eliminate the need for antiphase inputs, and this will also eliminate the need for a phase-splitter transformer as well. Transistors, being available in *n-p-n* or *p-n-p* forms, make possible a circuit arrangement which was impossible with valves because there was no valve 'equivalent' to a *p-n-p* transistor. *Figure 7.19*

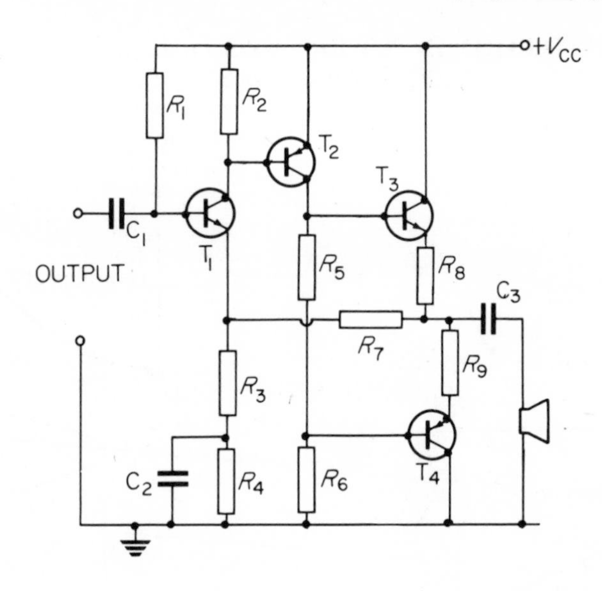

Figure 7.20

shows a modification of *Figure 7.18*, the transistors now being one of each type.

Such a mixture gives these circuits the name of 'complementary' and because of their symmetrical layout the system is called 'complementary symmetry'. What happens is that a single input signal drives one transistor on and the other off, so doing away with the need for antiphase drives. Again, the form of bias used (not shown for the same of clarity) will decide whether Class A or Class B operation is employed.

A practical form of the last circuit is shown in *Figure 7.20*. This has the further advantage that the split power supply used in the previous circuits is avoided by feeding the speaker load through a large value capacitor C_3 which is effectively a short-circuit to the signal currents but a d.c. block which does not disturb the d.c. conditions.

Transistors T_3 and T_4 form the complementary pair and these have a small forward bias derived from the current flowing in R_5 and R_6 due to the driver stage T_2. Resistor R_5 (which is often a thermistor or a diode) ensures that the voltage on the base of T_3 is slightly more positive than the voltage on the base of T_4; similarly, the voltage on the base of T_4 will be slightly more negative than the voltage on its emitter. T_3 and T_4 then pass the same quiescent current which is a few milliamperes above cut-off. R_7 provides some negative feedback to the emitter of T_1 and this tends to stabilise the quiescent potential at the junction of emitter resistances R_8 and R_9 at $V_{cc}/2$ volts. Notice the use of a *p-n-p* transistor for the driver stage T_2 so that it can be used 'upside-down' as it were.

(13) Can you say why the use of a thermistor or a diode is advantageous over a plain resistor in the R_5 position of this amplifier? (You might be able to cope with this problem better after reading the next section).

DISTORTION IN POWER AMPLIFIERS

When a transistor is operated over a wide range of its characteristics, distortion arises from the non-linearities inherent in semi-conductor devices. The input characteristic which relates base-emitter voltage to base current, has a characteristic of the form shown in *Figure 7.21(a)*,

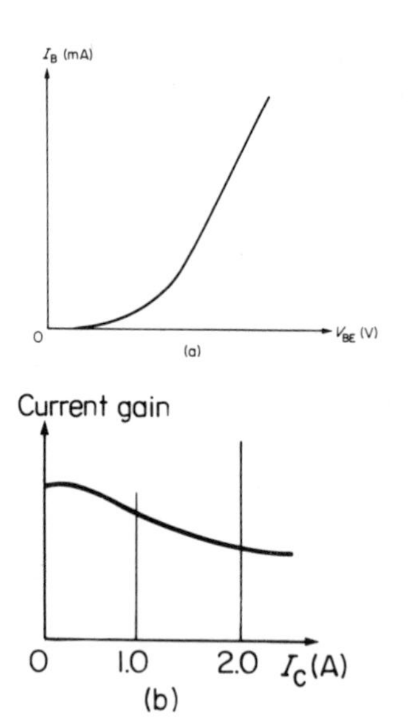

Figure 7.21

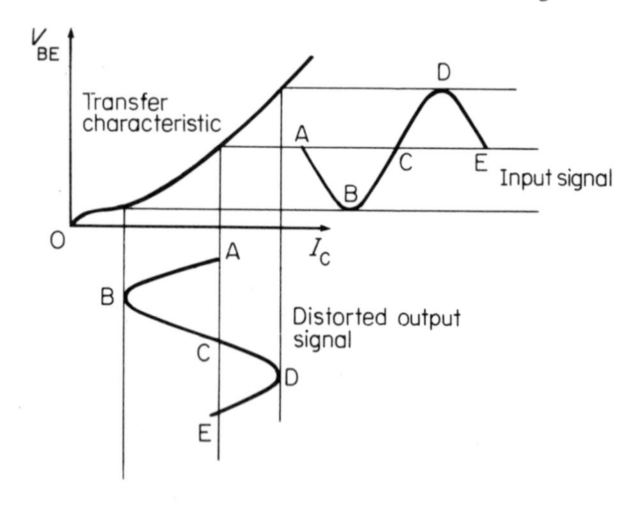

Figure 7.22

hence the input resistance of the transistor is not constant but varies with the amplitude of the signal. Also, the current gain of a transistor is dependent upon the collector current in a manner as shown in *Figure 7.21(b)*.

It is clear from this that if an amplifier is biased to a collector current of, say, 1 A, and the signal excursions range both positively and negatively from that point, then the positive excursion will get more amplification than the negative excursion.

Fortunately, the two effects illustrated in *Figure 7.21* tend to be self-cancelling so that the relationship between the base-emitter input voltage and the collector current is substantially linear. However, as *Figure 7.22* shows, the output wave is no longer the purely sinusoidal wave which we have assumed to be the input signal. A signal flattened on one side but peaked on the other indicates that it contains frequency components which were not present in the input signal. If these unwanted frequencies are harmonically related to the input signal frequency, the wave is harmonically distorted. It is beyond our present scope to analyse a wave into its harmonic components, but you can study the effect of various harmonics by reading (or revising) Section 9 in *Electrical and Electronic Principles 3*, (Butterworths).

A particular form of distortion arises in Class B amplifiers. The two transistors used in any push-pull amplifier should be matched in their dynamic characteristics as closely as possible and transistor pairs are available that are matched within specified limits in this way. What is known as 'crossover' distortion can arise in Class B amplifiers even though the transistors are perfectly matched.

Refer back to *Figure 7.14* and now suppose that we redraw that diagram with the transistors biased *exactly* to cut-off. The operating points P_1 and P_2 of the earlier figure now coincide where I_C is zero, and the two characteristic curves do *not* lie in a continuous straight line across the page as they did in the earlier figure. The situation is shown in *Figure 7.23*.

As the input signal passes from one to the other of the transistors, little base current flows until the input exceeds a value which carries it

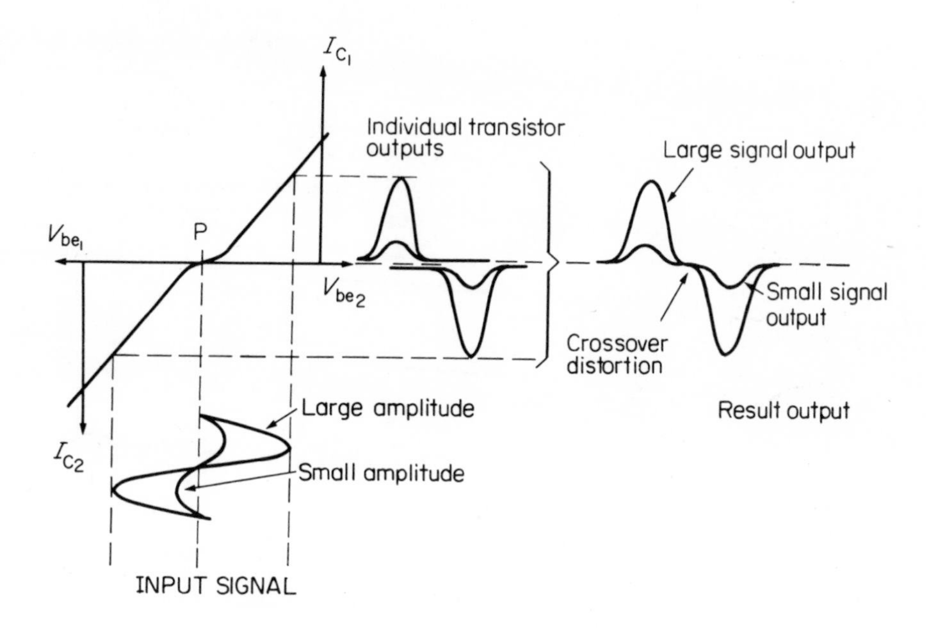

Figure 7.23

past the lower reaches of the transfer characteristics. Because I_C is almost directly proportional to base current, the resulting collector current will be very small until the input signal is sufficiently large. Hence, any one of the transistors is not fully turned on by the time the other is turned off.

As the figure shows, distortion takes place between the two half-cycles of each alternate cycle of output or during the 'cross-over' period. This trouble becomes worse as the signal amplitude decreases.

To overcome this type of distortion a slight forward bias is applied to the transistors so that a small collector current flows in the quiescent state. This effectively removes the curvature from the operation of the amplifier because as one transistor is being driven on to the non-linear region the other is already well into conduction and its collector current swamps out the small distorted current from the first. *Figure 7.14*, in fact, showed the proper biasing condition and you should carefully compare this diagram with *Figure 7.23* above.

This form of bias which is ahead of cut-off is strictly called Class AB since for *small* input signals the operation approximates to Class-A working. The method of course reduces the overall efficiency of the stage slightly.

PARASITIC OSCILLATIONS

It is possible for a power amplifier to generate oscillations at a very high frequency, possibly several MHz. When this unwanted oscillation mixes with the audio signals being amplified, harmonics are produced which can be detected as unpleasant background noise or interference. Additionally, when the amplitude of the oscillation is high, the static bias conditions of the circuit can be upset with resulting distortion of the wanted signals, overloading and, in extreme cases, actual damage to the transistors concerned. It is not always recognised immediately that an amplifier which is not properly functioning is probably suffering from such oscillations. The name *parasitic oscillation* is given to this kind of problem.

It is usually a simple matter to guard against and prevent such oscillations from being set up. A *small* value resistance may be placed directly in the base or collector lead as shown in *Figure 7.24(a)*, or a small value capacitor may be connected from collector to the earth rail as shown at (*b*). A ferrite bead is sometimes threaded on to the emitter lead as in diagram (*c*). The object of all these additions is to

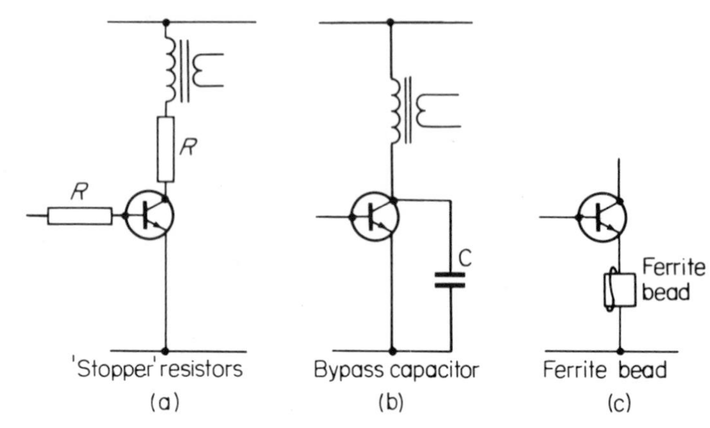

Figure 7.24

eliminate a resonant condition by the damping effect of resistance, bypass any such oscillations or shift the resonant frequency to some value beyond the capabilities of the transistor to amplify it. In a newly designed amplifier where the difficulty may persist, it could become necessary to revise the layout of components or try alternative types.

PROBLEMS FOR SECTION 7

(14) An amplifier has an efficiency of 55%. What class is it probably working in?

(15) The optimum load resistance of a certain output transistor is 150 Ω. What must be the turns ratio of the output transformer so that the transistor will be matched to a 3 Ω load?

(16) A single-ended Class A amplifier is biased to a quiescent collector current of 0.75 A and delivers a signal power of 4.2 W to a transformer coupled load. If the supply voltage is 16 V, calculate: (a) the efficiency, (b) the collector dissipation.

(17) A power amplifier has an efficiency of 24% and delivers 3.5 W to the output load. What d.c. power is taken from the supply?

(18) An amplifier has a collector dissipation of 2 W and an efficiency of 45%. What will be the signal power output from the amplifier?

(19) An amplifier with 30% efficiency is required to provide a 5 W output. Find the collector dissipation and the supply power.

If this amplifier (i.e. with the same dissipation) is used in a circuit having an efficiency of 70%, what will be the new output power?

(20) A Class A audio-frequency power amplifier takes a mean collector current of 15 mA. When a sinusoidal signal is applied to the amplifier the collector current swings between 28 mA and 2 mA and the collector voltage between 2 V and 18 V. If the V_{cc} supply is 10 V, calculate for this amplifier (a) the collector dissipation, (b) the signal power output, (c) the efficiency.

(21) List the advantages and disadvantages of Class B operation compared with Class A.

(22) A power transistor is supplied from a 15 V supply and feeds a 10 Ω load resistance through a 2:1 step-down transformer. The collector current varies sinusoidally between 650 mA and 50 mA. Determine (a) the output power, (b) the power taken from the supply, (c) the efficiency, (d) the collector dissipation. (C. & G.)

(23) A single stage Class A power amplifier uses a power transistor with its collector connected to a 15 V supply through the primary winding of a transformer. The secondary winding is connected to a 3 Ω load resistance. The primary to secondary turns ratio is 9:1. If the average collector current is 60 mA, the power in the load is 300 mW and the load current is sinusoidal, calculate: (a) the efficiency, (b) the collector dissipation, (c) the peak-to-peak voltage swing at the collector, (d) the maximum instantaneous collector current. (C. & G.)

(24) A Class A power amplifier uses a common-emitter transistor circuit. The transistor is rated at a maximum collector dissipation of 2.5 W and has a collector efficiency of 35% at full output. Calculate the maximum power which can be delivered to a load for a signal which is (i) always applied at a constant level; (ii) likely to vary, with long periods of no signal.

For *both* cases (i) and (ii) determine the d.c. collector current if the d.c. collector voltage has a value of 15 V and the load is transformer coupled. (C. & G.)

(25) Explain why the internal dissipation of a Class A amplifier rises when the input signal is reduced in amplitude.

A Class A power amplifier has a 4 Ω transformer coupled load. With the maximum load power of 2 W the efficiency is 35%. Assuming the output to be sinusoidal (i) calculate the internal dissipation when the output power is at its maximum; (ii) plot a curve showing the variation of internal dissipation with load voltage as the input signal is reduced to zero. (C. & G.)

(26) The collector characteristics of a *n-p-n* power transistor are given below:

V_{ce} (V)	Collector current I_C (A) for base current I_b mA					
	I_b = 10	20	30	40	50	60
2	0.45	0.87	1.25	1.60	1.92	2.40
20	0.50	0.97	1.40	1.80	2.16	2.60

The characteristics may be assumed straight between the given limits. The transistor is so mounted that the maximum permitted collector dissipation is 18 W. It is used in a Class A transformer coupled amplifier circuit with a supply V_{cc} of 12 V.

(a) Plot the characteristics and draw in the maximum dissipation curve.

(b) Select a suitable operating point and drawn an a.c. load line.

(c) Estimate the output power when a sinusoidal signal of 25 mA peak amplitude is applied to the amplifier, and the efficiency under this condition.

8 Stabilised power supplies

Aims: At the end of this Unit section you should be able to:
Identify the requirements for maintaining a constant voltage output across a load.
Understand the principles of shunt and series stabilisation.
Sketch the block and circuit diagrams of simple series stabilised power supplies using comparator methods.
Explain the operation of such stabilised power supplies.

Direct current power supplies for electronic equipment may, in many cases, be obtained from cells or batteries. Pocket calculators, for example, are nearly always designed to operate from a number of dry cells and in a large number of cases, provision is made for these cells to be periodically recharged, so avoiding the inconvenience of replacement for considerable periods. Flashlamps and portable transistor radio receivers along with car radio and cassette players are other familiar examples where battery supplies are used.

However TV receivers are not usually operated by batteries (apart from those designed for this purpose) and neither would it be very economical or particularly practicable to operate, say, large computer systems or high-power transmitters, or even the domestic hi-fi equipment, from battery supplies. In all these cases, the power is obtained from rectified a.c. mains supplies.

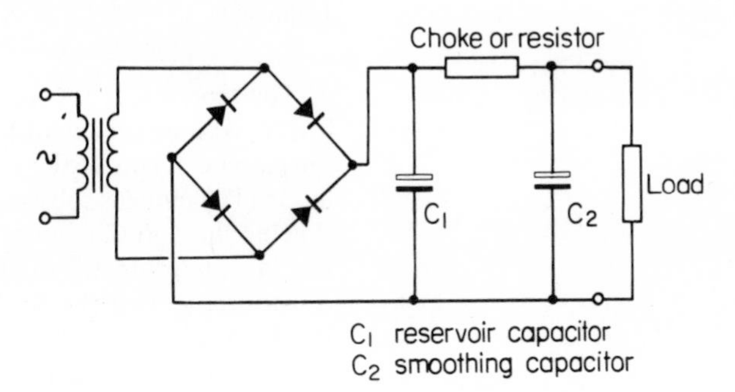

Figure 8.1

Rectification of the alternating supply is a relatively simple matter. Full-wave or bridge rectifier circuits used in conjunction with suitably wound transformers will provide us with a direct voltage output, and if this output is smoothed and filtered by a choke-capacitor or resistance-capacitor circuit, a reasonably steady output is available. Such a basic power supply system is shown in *Figure 8.1* and for many pieces of electronic equipment such a supply is perfectly adequate. For more demanding conditions however, for example, in supplying d.c. amplifier or stable oscillators, circuits of this sort are unsuitable. The main reasons for this are:

(i) The output voltage changes if the mains input voltage varies;
(ii) The output voltage changes if the load conditions vary;
(iii) The ripple component in the output which, for 50 Hz supplies is at a frequency of 100 Hz, increases as the current demands of the load increases.

Such supplies are said to be *unstabilised* against these (usually inevitable) variations. A *stabilised* or regulated power supply, on the other hand, is designed to reduce ripple and ripple variations to a minimum and to provide a stable output voltage regardless of mains voltage or load current variations. In other words, we want a circuit which behaves as closely as possible to the theoretical constant-voltage generator which has negligible internal resistance.

There are two types of regulated power supplies—shunt regulators and series regulators. We will investigate these in turn.

SHUNT REGULATION

The simplest form of shunt regulation is the single zener diode connected as shown in *Figure 8.2.* The regulator element (the zener) is wired in parallel with the load and is fed current by way of series resistor R_S. The elementary theory of zener stabilisation was covered in the Electronics 2 course.

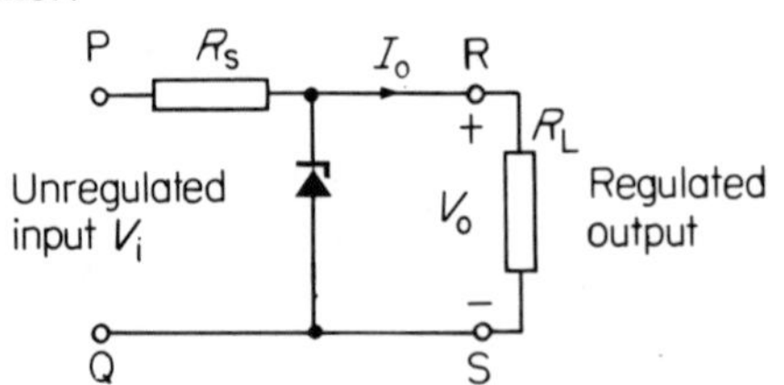

Figure 8.2

The zener diode is connected into the circuit so that it is reverse-biased. At a particular voltage for which the diode has been designed, breakdown occurs at the junction and thereafter the voltage across the diode (V_z) remains substantially constant, irrespective of the current flowing through the diode. This reverse current flow would be destructive in an unprotected diode, but here it is limited to a safe value by resistor R_S so that the power dissipation of the zener is not exceeded. Provided that the voltage across the zener does not fall below the breakdown value, the diode behaves as a current reservoir.

Referring to *Figure 8.2*, it is not difficult to understand how the zener diode provides a constant output voltage V_o at terminals R and S in spite of variations in either the supply voltage V_i at terminals P and Q or in the load current I_o flowing through R_L. Suppose the d.c. input voltage at P-Q increases for some reason. Then the current through the zener diode increases but as the voltage across its terminals remains constant, the increase in voltage appears across R_S. Conversely, if the d.c. input at P-Q decreases, the zener diode surrenders the extra current and the voltage across R_S falls. Hence the input variation is *absorbed by resistor* R_S and the wanted output voltage at R-S remains constant.

Suppose now the load current I_o increases for some reason. The zener current will decrease by the same amount. Similarly, if the load current decreases, the zener current will increase by the same amount.

This time the zener takes up any excess current and sheds any current difference demanded by the load, so acting as a current reservoir while maintaining a constant voltage at the output terminals. General purpose zener diodes are readily available in a range of breakdown voltages running from about 2.5 V up to some 200 V, with power ratings ranging from 400 mW up to 20 W and more.

SERIES REGULATION

In the series form of regulator, the control element is connected in series with the supply lead to the load. If the input voltage V_i or the load current I_o change, the voltage across the control element will adjust itself in such a way that the output voltage will remain constant.

A basic form of series regulator is shown in *Figure 8.3.* This circuit is essentially a current amplifier used in conjunction with a zener diode which acts as a fixed voltage reference V_{ref}. Strictly, we have a shunt regulator (the zener diode) acting through a series element (the current amplifier).

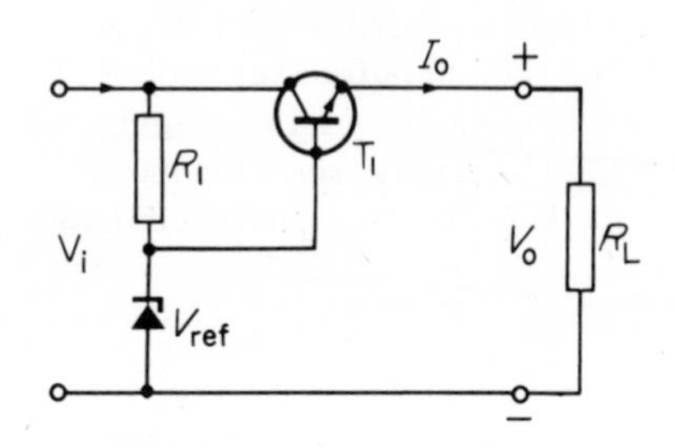

Figure 8.3

The power transistor T_1 is capable of supplying a relatively large current. Its base is held at a constant voltage by the zener diode, supplied with current through resistor R_S exactly as we have already described. This transistor then acts as a current amplifier, so enabling the circuit to provide much higher currents than can be handled by the zener diode alone. If, for example, the current gain of the transistor is 50, then a base current of 20 mA will provide an output current of 1 A. Such a base current is now the effective zener load.

The circuit regulates as follows: suppose the base-emitter voltage is 0.75 V when the transistor is delivering a current of 1 A, then the output voltage will change by 0.75 V when the current drawn from the circuit changes by 1 A. Hence the resistance of the supply as seen at the output terminals is $\delta V_o/\delta I_o = 0.75/1 = 0.75\Omega$.

Although such a circuit is often adequate for supplying relatively constant load currents at a fixed voltage, some degree of sophistication is called for in more stringent cases.

COMPARATOR METHODS

In *Figure 8.4* an *error amplifier* is used to *compare* the output voltage V_o with a reference voltage V_{ref} and controls the current flowing in the series regulator in such a direction that the error is minimised.

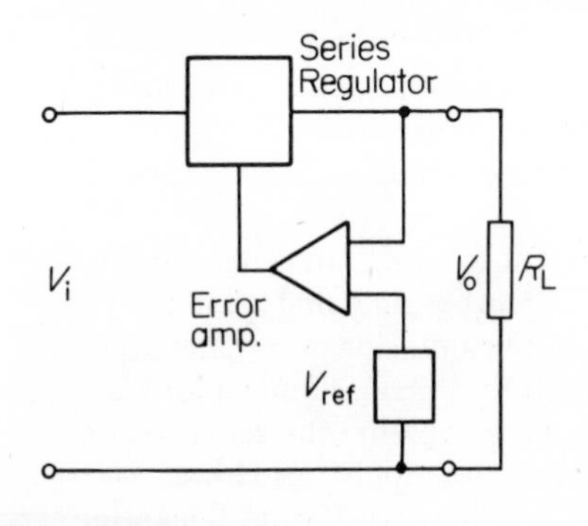

Figure 8.4

A practical form of this circuit is shown in *Figure 8.5.* Transistor T_1 is the series regulator, transistor T_2 with collector load R_2 is the error amplifier. The emitter of T_2 is held at a constant voltage determined by the zener diode fed, in the usual way, through R_1. Resistors R_3 and R_4 feed back a portion of the output voltage to the base of T_2 and are selected so that, in addition, the base of T_2 is positive with respect to its emitter. Hence T_2 is conducting and there is a voltage drop across its collector load R_2. The output at the collector feeds directly to the base of T_1.

Suppose now an output voltage change occurs. This change is sensed at the junction of R_3 and R_4 and hence at the base of T_2. Since the emitter potential is fixed, the base variation will either increase the current through T_2 or decrease it, according as the output voltage variation is an increase or a decrease respectively. The variation, in other words, is compared with the zener reference, and the transistor senses in which direction its base potential has shifted relative

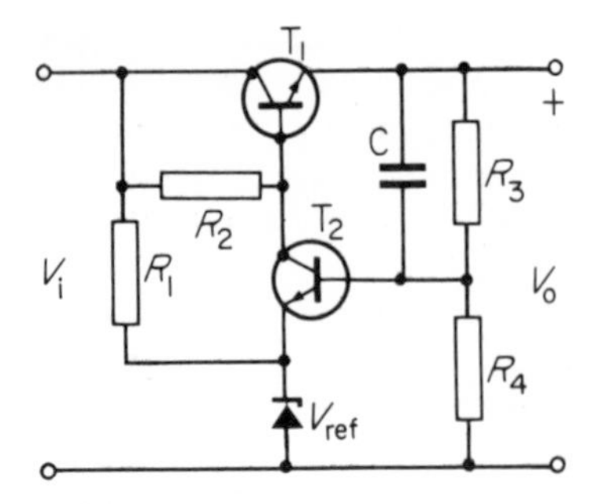

Figure 8.5

to that reference. The change is amplified by T_2 and is fed to the base of T_1 where it adjusts the output current flowing through T_1 in such a way that the original output variation is minimised. You will probably spot that this is an example of negative-feedback, and the circuit can be analysed in terms of negative-feedback theory. This is, however, beyond your immediate course requirements.

In circuits of this sort, the gain of T_2 should be as large as possible, and T_1 must be capable of dealing with the full output current and generally with high power. The voltage drop across T_1 must also be large enough at all times to ensure that it conducts at all times. You may wonder about the capacitor C connected across R_3. This is included to enable the circuit to cope with rapid voltage variations, for R_1 is then effectively short-circuited by C and the full output variation is imposed on the base of T_2. For slow output variations, the output change is attenuated by the potential divider R_3-R_4; since the base and emitter of T_2 are practically at the same potential; the attenuation is in fact

$$\frac{V_o}{V_{ref}} = \frac{R_3 + R_4}{R_4}$$

From this relationship it can be seen that

$$V_o = \left[\frac{R_3 + R_4}{R_4}\right] V_{ref} = \left[\frac{R_3}{R_4} + 1\right] V_{ref}$$

so that V_O can be adjusted by manual variation of the ratio R_3/R_4.

A circuit with adjustable output voltage control is shown in *Figure 8.6.* Potentiometer VR_1 replaces the fixed divider chain and permits a limited range of control, but the output voltage can never be less than that of the reference. If it is required to control the output voltage down to zero, a negative supply rail has to be provided.

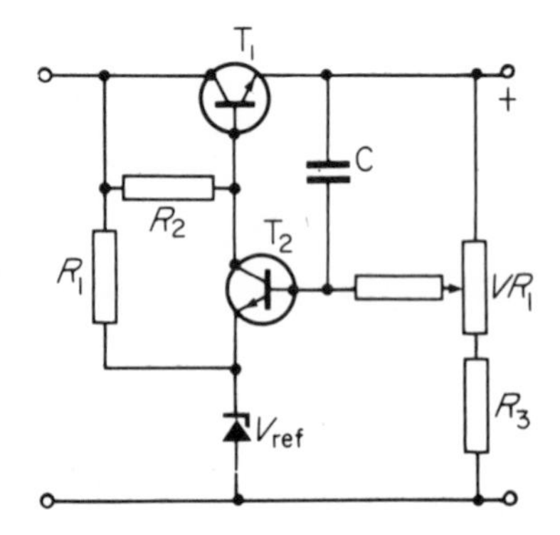

Figure 8.6

(1) Why cannot the output voltage be reduced below the reference level?

(2) What would happen to the output voltage, if anything, if it did drop bélow the reference level?

Example (3) *Figure 8.7* shows a series stabilised regulator using two transistors, T_2 and T_3, in the amplifier feedback path. Deduce how this circuit operates to maintain a constant output voltage level.

This circuit looks rather different from the previous one. The zener diode, for example, is on the 'other side' of T_1 and its feed resistor R_4 is returned to the negative rail. However, taking the circuit a step at a time soon reveals that there is quite a lot we will recognise. Clearly, it is immaterial whether the zener has its series resistor on the positive side of the supply or the negative. The zener will break down at its rated voltage V_z and thereafter it will behave as a constant voltage source. Ignoring R_3 for the moment, the output voltage will be the sum of V_z and the base-emitter voltage of T_3 (about 0.6 V). So the zener rating

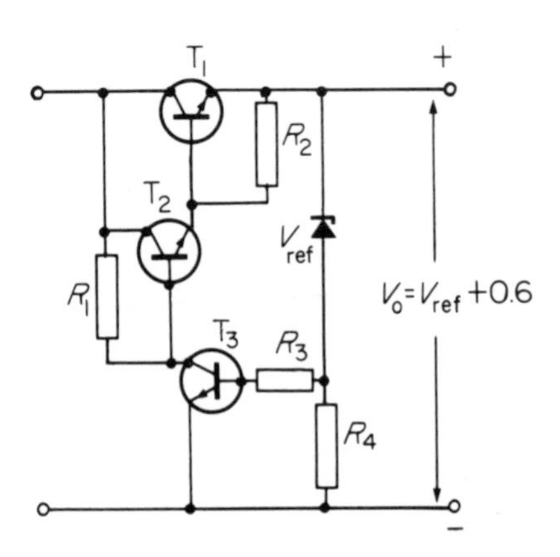

Figure 8.7

determines the order of output voltage we are going to get.

We have already noted that the resistor in series with a zener takes up any voltage changes occurring across the combination; hence if the output voltage tends to change, the potential across R_4 will change accordingly and the base input to T_3 also changes. The output of T_3 drives the base of T_2; this transistor is in emitter-follower connection and the load impedance output at the emitter controls the series regulator T_1 forming the emitter load of T_2.

Check the polarity variation round the circuit for yourself and verify that the control exercised by T_1 is in the proper direction to reduce the original output voltage change.

Resistor R_2 is included to shunt away the collector-base leakage current of T_1, while R_3 is selected to limit the base current of T_2 in the event of a fault developing.

PROTECTION

There is no inherent short-circuit protection for the series regulator as there was for the shunt circuit. If the output terminals of the series regulator were accidentally short-circuited or if an excessive current was inadvertently drawn by the load, the very large resulting current would, in all probability, cause excessive power dissipation in the control transistor and burn it out.

Protection can be provided by additional circuitry as shown in *Figure 8.8*. When an excessive current flows in the current sensing resistor, the voltage developed across it becomes great enough to switch on the overload detector. This then takes control away from the error amplifier and biases the series regulator back to a safe current condition, usually arranged to be just above the normal maximum current limit.

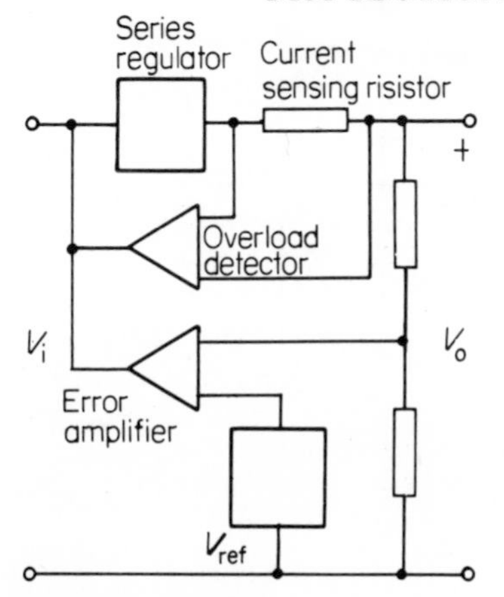

Figure 8.8

Again, a practical form of such a system is shown in *Figure 8.9*. This appears even more complicated, but we have the circuit of *Figure 8.7* with one additional transistor, T_4. This forms the overload detector. When the current drawn by the load exceeds a certain level, the voltage across R rises sufficiently from the base-emitter voltage of T_4 to exceed its switch-on value of 0.6 V. T_4 then conducts and diverts base current from T_3, hence limiting the output current to a safe, preset value. The point at which T_4 switches on (in terms of the voltage across R) is decided by the setting of VR_1 so a range of current limiting values is possible. The current sensing resistor R is usually of 1 to 2 Ω resistance, with the current setting control a few hundred ohms.

This is as far as you need to go into the study of basic voltage regulator systems. We have covered only the most elementary methods of such control, but the principles involved are common to the most sophisticated circuits.

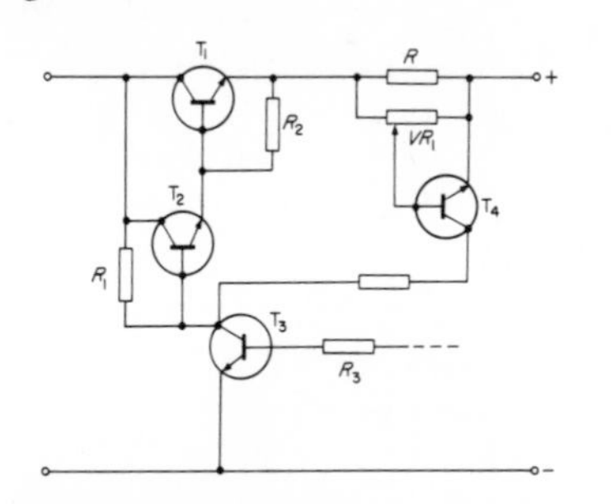

Figure 8.9

PROBLEMS FOR SECTION 8

(4) Differentiate between an unstabilised and a stabilised d.c. power supply.

(5) A 20 V stabilised supply is required from a 40 V unstabilised d.c. source. A 20 V zener diode is to be used for this purpose, having a power rating of 1.5 W. Find the required value of the series resistor.

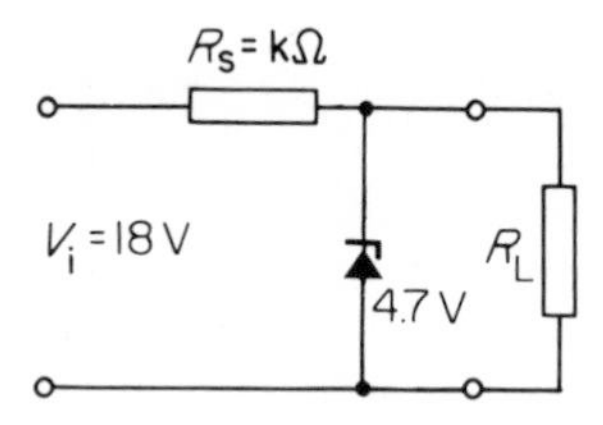

Figure 8.10

(6) A zener diode is to provide a 16 V stabilised output from a 20 V unstabilised supply. The load resistor R_L = 2 kΩ and the zener current I_Z = 8 mA. What series resistor is required and what power will be dissipated in it.

(7) In the circuit of *Figure 8.10*, what is the maximum power dissipation in the zener diode? What is the greatest load current that can be drawn from this circuit if the minimum zener current is 3 mA?

(8) The slope resistance of the zener of the previous example is 20 Ω. What will be the change in the no-load output voltage when the input voltage falls to half its present value?

(9) State the requirements for maintaining a constant voltage output across a load. Explain the operation of the power supply unit shown in *Figure 8.11*.

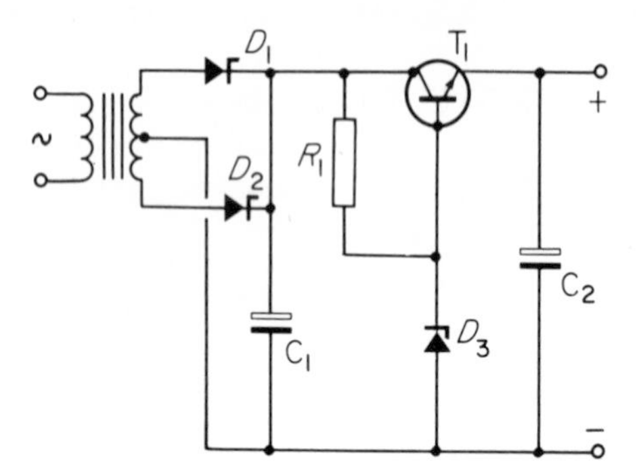

Figure 8.11

Solutions to problems

SECTION 1

(1) (a) one, (b) 0.033 s, (c) 10 V, (d) 0.005 s
(3) 500 Ω; 60 mA
(4) (a) 100 V; (b) 40 V; (c) 0 V
(7) 4 kΩ; (a) 2.5 mA; (b) 25 V/s
(8) See *Figure A.1.*
(12) (a) 320 ms, 1 : 2.2; (b) 13 V, –3.5 V; (c) 50 μs, 100 μs
(13) (a) 0.8 μs, (b) 2.4 μs; (c) 5 V; (d) 102 μs
(15) About 0.027 s
(17) (a) 0.05 mA; (b) 12.5 V/s; (c) 6.25 V/s
(19) For a differentiator, $f \leqslant 5$ Hz; for an integrator, $f \geqslant 500$ Hz.
(21) (i) 52.3 ms; (ii) 23.2 ms

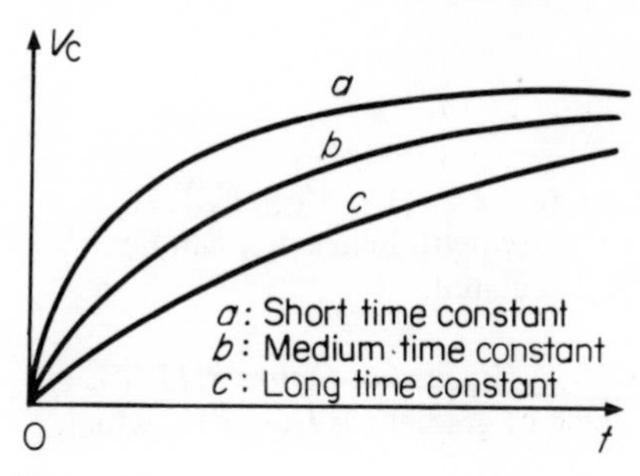

Figure A1

SECTION 2

(3) 3 dB = 20 log (voltage ratio). For 3 dB *reduction*, voltage ratio = antilog –3/20 = 0.707.

(4) (b) is a wideband or video amplifier; (c) is a tuned or selective amplifier.

(6) (a) 1.25 mA; 31.

(7) No, because I_b will shift with output-current variations.

(8) To prevent feedback at the signal frequencies. (*see* Section 5).

(11) 143 Hz.

(12) 0.84 μF.

(16) Depends upon the frequency range. If only d.c. and *very* low frequencies were involved, there would be no point in bypassing R_2.

(17) For example: 'An input signal V_i will be amplified 100 times as much as any temperature induced voltage at the input'.

(18) 40 μA; 32.3.

(20) (a) direct; (b) the whole; (c) low.

(21) 40 dB, 32 dB, 14 dB, –6 dB; 100, 5, 4, 0.001

(22) (a) 20; (b) 400; (c) 26 dB.

(23) 8.15A.

(24) 40 times.

(27) 72 kΩ.

(28) (a) 1.2 mA; (b) 1.25 mA. The bias arrangement maintains I_C at a substantially constant level even when the transistor is changed for one with twice the static gain figure.

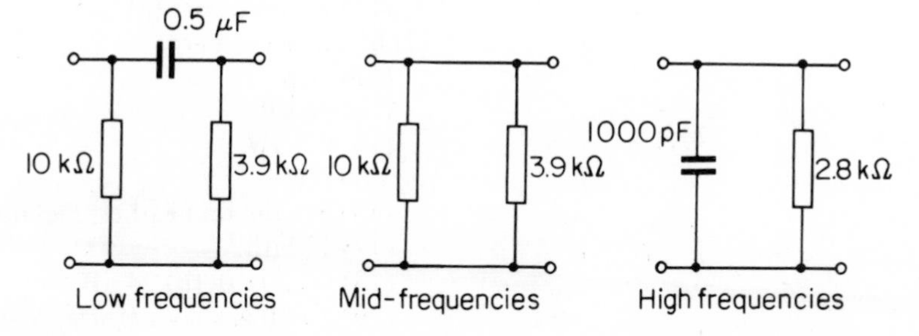

Figure A2

(29) About 17 kHz. This is a tuned selective amplifier, most likely an intermediate-frequency amplifier in a superheterodyne.

(30) About 2 kHz. Most likely a transformer-coupled amplifier.

(31) (a) 4 kΩ; (b) 85.

(32) $A_v = 58$; $A_i = 48$; $A_p = 2784$; (approx).

(33) 31.8 kHz.

(34) The coupling equivalent circuits are shown in *Figure A2.* The bandwidth lies between 23 Hz and 56 kHz.

(36) (a) 7.9 mV; (b) 25 μW.

(37) 100 μV per °C

SECTION 3

(2) Doing the substitution we get $g_{fs} = g_{fso}\ [1 - V_{GS}/V_p]$. I_{DSS} and V_p are easily found by direct measurement, hence g_{fso} can be found. Then g_{fs} for any V_{GS} can be calculated.

(5) 12.5 kΩ

(7) The gradient at $V_{GS} = 0$ is $2I_{DSS}/V_p$. From *Figure 3.11* it is at once clear that the tangent of the angle of gradient is $I_{DSS}/\frac{1}{2}V_p$ which gives the required proof.

(8) $r_d = 6.25$ kΩ; $g_{fs} = 6$ mS, $\mu = 37.5$.

(13) 3 mA, assuming V_{DS} remains constant.

(14) 66.7 kΩ.

(15) $\mu = 44$; $g_{fs} = 3.3$ mS, $r_d = 13.3$ k.

(16) (a) thermionic valves; (b) depletion; (c) *n*- or *p*-channel construction; (d) zero; (e) source to drain.

(17) $\mu = 10$; $A_v = 16.7$

(18) About 15 kΩ

(19) (a) –0.5 V; (b) 1.48 mS; (c) 470 Ω; (d) 6750 Ω.

(20) $r_d \simeq 6.67$ kΩ; $g_{fs} \simeq 1.7$ mS

(21) $R_1 = 3$ MΩ; $R_2 = 1$ MΩ. $A_v \simeq 7.5$.

(22) About 2.0 mS.

(23) 460 kHz; about 23. Note that we cannot use the approximate formula for voltage gain $A_v = -g_{fs}R_L$ here since R_L (=R_D) = 66.7 kΩ and this is much greater than r_d.

(24) (i) 280 pF; (ii) 294; (iii) 135.

SECTION 4

(2) No, because of the severe loss of gain.

(4) Less. There are fewer carriers crossing the junction barriers.

(6) 10^{-4} mW.

(7) S/N will be unchanged at 40 dB since no more noise is added.

(8) The signal is constant so the noise must be reduced by 15 dB. This is a ratio of 31.6 times ∴ the bandwidth is 10 kHz/31.6 = 316 Hz.

(11) Examples are (a) hum; (b) hiss; (c) ignition interference; (d) lightning discharge.

(12) 90 μV.

(13) 26 dB.

(14) 0.5 μW.
5.69 mW.

(15) See the text earlier, *Example (3).*

(17) 120 dB.

(18) (a) 30 dB; (b) 24 dB.

(20) 10 dB is a poor figure. A good figure would be 3-4 dB.

(21) (a) 12500 pW; (b) 1250 pW; (c) 11250 pW. If the bandwidth is halved the noise power will be halved.

(22) (a) –84 dBm; (b) –85.3 dBm. 4.02.

SECTION 5

(3) 9.93.

(4) 100.

(5) 300.

(6) 0.008.

(7) β would be unity, hence $A_v' = A_v/(1 + A_v)$ which is very close to 1. Such a circuit is possible and will be discussed in due course.

(9) No. n.f.b. reduces the noise output in exactly the same ratio as the gain, thus for a given signal input S/N ratio at the output is unaffected. We cannot apply the arguments that we did about harmonic distortion reduction to this case.

(10) The respective gains become 90.9, 98, 99, 90.9. *Figure A.3* shows the new response curve.

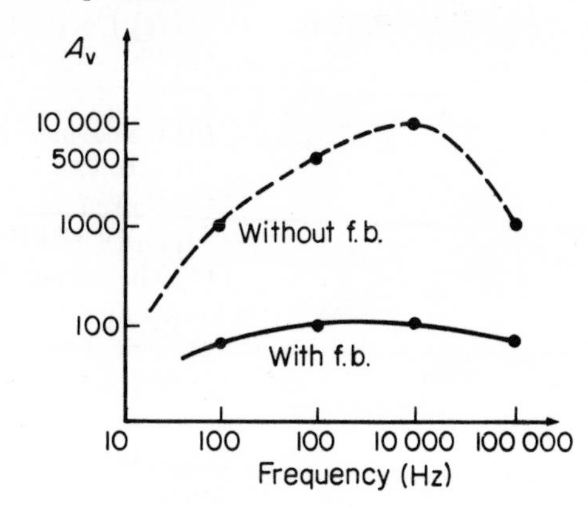

Figure A3

(13) 10 dB.

(14) $\beta = 0.08$. Bandwidth is 20.2 Hz to 1.237 MHz.

(15) No, the base bias would be upset.

(16) If R_2 is disconnected, *n.f.b.* is still applied to T_1 by the signal voltage developed across the unbypassed R_1 (the emitter-follower effect).

(17) The table should read:

A_v	β	βA_v	A_v'
100	–0.1	–10	9.1
1500	–0.2	–300	48.4
154	–0.05	–7.7	17.7
25	–0.25	–6.25	3.45

(19) (a) True; (b) True; (c) False.
(a) False; (b) True; (c) True.

(21) (a) 66.7; (b) ∞; (c) 25; conditions represent positive f.b, instability, negative f.b. respectively

(22) Positive f.b. $\beta = 0.004$

(23) 800.

(24) 91.7 to 90.0.

(25) 48.8; 47.6.

(26) 0.043.

(27) 29.7 dB.

(28) R_i is doubled, R_o is halved.

(29) 30.7 dB; 0.03 times

(30) $\beta = 0.049; A_v = 2220$. (You need simultaneous equation for this problem).
(31) 49.7.
(32) (a) 19.05; (b) 18.7; (c) 0.98.
(33) 10; 50 kHz.
(34) 0.77.
(36) 20 dB.
(37) The gains with f.b. are 10.3, 11.4, 11.7 and 11 dB.
(38) 10 pF. Capacitive impedance is greater with a *smaller* capacitance
(39) 0.041; 38.5.

SECTION 6

(1) (a) $1/\beta$; (b) frequency.
(2) Relaxation.
(3) 90°
(4) It becomes smaller, reducing to zero at 90° shift.
(6) 250 Hz.
(7) Halved.
(9) (i) 13.18 kHz; (ii) About 1.82 to 1.
(11) The output from T_2 collector depends only upon the transistor response time. At the other collector, the coupling capacitor has to charge from almost zero potential to V_{cc} through the collector load which takes a finite time determined by the time constant of these components.
(13) (a) amplitude; (b) low; (c) frequency; (d) one-quarter as great; (e) lowest; (f) $+V_{cc}$.
(14) C_2R_3.
(16) 2.9 kHz.
(17) (a) False; (b) True; (c) False.
(18) Volts drop across 1.2 kΩ = 1.2(3.21 + 0.05) = 3.91 V. Therefore $V_{ce} = 9 - 3.91 = 5.09$ V.
(20) 1.25 kHz.
(21) $128 = 2^7$, therefore 7 stages are required.

SECTION 7

(2) 15 W; 20 W.

(3) (a) $$\left[\frac{\hat{I} - I_Q}{\sqrt{2}}\right]^2 R_L' = \frac{[\hat{I} - I_Q]^2}{2} R_L'$$

(b) $$\frac{\hat{V} - \check{V}}{2\sqrt{2}}^2 \frac{1}{R_L} = \frac{[\hat{V} - \check{V}]^2}{8} \frac{1}{R_L'}$$

(4) In this case $\hat{V}$ would be V_{cc} and not $2V_{cc}$; hence

$$\eta = \frac{V_{cc}.2I_Q}{8V_{cc}.I_Q} = \frac{1}{4}.$$

(8) (a) $\hat{P}_L = \dfrac{V_{cc}^2}{4R_L'}$; (b) $\hat{P}_{dc} = \dfrac{V_{cc}^2}{\pi R}$

(9) 1.6 : 1 each half.
(10) (a) zero; (b) difference.

(12) Common-collector or emitter-follower stages.

(13) A thermistor or diode resistance changes with temperature and so compensates for temperature variations if it is mounted close to the power transistor (s).

(14) Class AB or B.

(15) 7 : 1.

(16) (a) 35%; (b) 7.8 W.

(17) 14.58 W.

(18) 1.63 W.

(19) 16.7 W; 11.7 W; 27.3 W.

(20) (a) 150 mW; (b) 52 mW; (c) 34.7%.

(21) Advantages: Higher output power for a given V_{cc}.
Negligible collector dissipation in the absence of a signal.
Higher efficiency.

Disadvantages: V_{cc} supply must have good regulation.
Self bias not possible for exact Class B.
Harmonic distortion higher.

(22) (a) 1.8 W; (b) 4.5 W; (c) 40%; (d) 2.7 W.

(23) (a) 33.3%; (b) 0.6 W; (c) 24 V; (d) 96 mA.

(24) (i) 1.346 W; (ii) 0.875 W; (i) 77 mA; (ii) 166 mA.

(25) 3.7 W.

(26) Your operating point should be at V_Q = 12 V (V_{cc}), I_Q = 1.5A, with the load line tangential at this point to the maximum dissipation curve. This gives a base bias of about 35 mA.

The output power is then about 3.3 W and the efficiency about 18%.

SECTION 8

(1) Transistor T_2 would turn off.

(5) 267 Ω.

(6) 46 Ω; 0.35 W.

(7) 62.5 mW; 10.3 mA.

(8) 180 mV.